Methods and Protocols in Food Science

Series Editor
Anderson S. Sant'Ana
University of Campinas
Campinas, Brazil

For further volumes:
http://www.springer.com/series/16556

Methods and Protocols in Food Science series is devoted to the publication of research protocols and methodologies in all fields of food science.

Volumes and chapters will be organized by field and presented in such way that the readers will be able to reproduce the experiments in a step-by-step style. Each protocol will be characterized by a brief introductory section, followed by a short aims section, in which the precise purpose of the protocol will be clarified.

Basic Protocols in Predictive Food Microbiology

Edited by

Verônica Ortiz Alvarenga

Faculty of Pharmacy, Department of Food, Federal University of Minas Gerais, Belo Horizonte, MG, Brazil

Humana Press

Editor
Verônica Ortiz Alvarenga
Faculty of Pharmacy
Department of Food
Federal University of Minas Gerais
Belo Horizonte, MG, Brazil

ISSN 2662-950X ISSN 2662-9518 (electronic)
Methods and Protocols in Food Science
ISBN 978-1-0716-3412-7 ISBN 978-1-0716-3413-4 (eBook)
https://doi.org/10.1007/978-1-0716-3413-4

This Humana imprint is published by the registered company Springer Science+Business Media, LLC, part of Springer Nature.
The registered company address is: 1 New York Plaza, New York, NY 10004, U.S.A.

Preface to the Series

Methods and Protocols in Food Science series is devoted to the publication of research protocols and methodologies in all fields of food science. The series is unique as it includes protocols developed, validated and used by food and related scientists, as well as theoretical basis are provided for each protocol. Aspects related to improvements in the protocols, adaptations and further developments in the protocols may also be approached.

Methods and Protocols in Food Science series aims to bring the most recent developments in research protocols in the field as well as very well-established methods. As such, the series targets undergraduate, graduate and researchers in the field of food science and correlated areas. The protocols documented in the series will be highly useful for scientific inquiries in the field of food sciences, presented in such way that the readers will be able to reproduce the experiments in a step-by-step style.

Each protocol is characterized by a brief introductory section, followed by a short aims section, in which the precise purpose of the protocol is clarified. Then, an in-depth list of materials and reagents required for employing the protocol is presented, followed by a comprehensive and step-by-step procedures on how to perform that experiment. The next section brings the do's and don'ts when carrying out the protocol, followed by the main pitfalls faced and how to troubleshoot them. Finally, template results are presented and their meaning/conclusions addressed.

The *Methods and Protocols in Food Science* series will fill an important gap, addressing a common complaint of food scientists, regarding the difficulties in repeating experiments detailed in scientific papers. With this, the series has a potential to become a reference material in food science laboratories of research centers and universities throughout the world.

Campinas, Brazil ***Anderson S. Sant'Ana***

Preface

Predictive microbiology is a tool for modeling and simulation based on microbial responses to environmental factors. Several models can be developed considering the problem, the response variable, and the data collection. From the predictive models, it is possible to study the time for the formation of toxins or compounds that limit the shelf life of food, multiplication, survival, or inactivation of spoilage microorganisms or pathogens.

Predictive models produce useful information for decision making because they help to establish critical boundaries based on microbial behavior under changing environmental conditions. The data generated from the predictive models provide essential information for elaborating HACCP plans, food safety metrics, and Quantitative Microbiological Risk Analysis (QMRA). Also, predictive models can be applied to evaluate food shelf life, food formulation modification, and evaluation of the effect of processing steps on microorganisms.

This book aims to comprehensively introduce methods and procedures related to the design of the experiments, data collection, and data analyses for elaborating predictive models. All chapters include introductions to the respective topic, the most common protocols, a list of materials and reagents, readily reproducible stepwise laboratory protocols, and the hallmark of the MeFS series.

This book combines well-established protocols and procedures used by many laboratories in academia and industry as comprehensibly as possible. It introduces the broad field of Predictive Microbiology in Foods to young researches, i.e., graduate students, postdoctoral associates, and all researchers who are either still at the beginning of their academic career or scientists in search of new challenges in a new field hitherto unfamiliar to them.

The book is divided into three main sections. The initial segment covers general techniques for designing and gathering data to create predictive models. The second part focuses on the development of predictive models. Finally, the third section explores the topic of microbial behavior from several approaches. In addition, the book supports researchers looking to expand the range of their investigations in the predictive microbiology field.

Belo Horizonte, MG, Brazil — *Verônica Ortiz Alvarenga*

Contents

Contributors

VERÔNICA ORTIZ ALVARENGA • *Faculty of Pharmacy, Department of Food, Federal University of Minas Gerais, Belo Horizonte, MG, Brazil*

ZAFEIRO ASPRIDOU • *Department of Food Science and Technology, School of Agriculture, Faculty of Agriculture, Forestry and Natural Environment, Aristotle University of Thessaloniki, Thessaloniki, Greece*

ISABELLA BASSOTO XAVIER • *Department of Food, Faculty of Pharmacy, Federal University of Minas Gerais, Belo Horizonte, MG, Brazil*

ARACELI BOLÍVAR • *Department of Food Science and Technology, UIC Zoonosis y Enfermedades Emergentes ENZOEM, ceiA3, Universidad de Córdoba, 14014, Córdoba, Spain*

LUMA MOURA BRITO • *Department of Food, Faculty of Pharmacy, Federal University of Minas Gerais, Belo Horizonte, MG, Brazil; Instituto de Tecnologia de Alimentos e Bebidas, CTI SENAI, Belo Horizonte, MG, Brazil*

MICHAEL CALLANAN • *Department of Biological Sciences, Munster Technological University, Cork, Ireland*

BRUNO AUGUSTO MATTAR CARCIOFI • *Federal University of Santa Catarina, Department of Chemical and Food Engineering, Center of Technology, Florianópolis, SC, Brazil*

JEAN CARLOS CORREIA PERES COSTA • *Department of Food Science and Technology, UIC Zoonosis y Enfermedades Emergentes ENZOEM, ceiA3, Universidad de Córdoba, 14014, Córdoba, Spain*

ALICE KARINE DA SILVA • *Department of Food, Faculty of Pharmacy, Federal University of Minas Gerais, Belo Horizonte, MG, Brazil*

MÍRIAN PEREIRA DA SILVA • *Department of Food Technology, Federal University of Viçosa, Viçosa, Minas Gerais, Brazil*

GLÁUCIA MARIA FALCÃO DE ARAGÃO • *Federal University of Santa Catarina, Department of Chemical and Food Engineering, Center of Technology, Florianópolis, SC, Brazil*

NATAN DE JESUS PIMENTEL-FILHO • *Center for Natural Sciences, Federal University of São Carlos, Buri, São Paulo, Brazil*

MIRIAN CRISTINA FEITEN • *Food Engineering Department, State University of Maringá, Paraná, Brazil*

PABLO S. FERNANDEZ • *Departamento de Ingeniería Agronómica, Campus de Excelencia Internacional Regional "Campus Mare Nostrum", Instituto de Biotecnología Vegetal, Escuela Técnica Superior de Ingeniería Agronómica, Universidad Politécnica de Cartagena, Cartagena, Spain*

ALBERTO GARRE • *Food Microbiology, Wageningen University & Research, Wageningen, The Netherlands*

LEONIDAS GEORGALIS • *Departamento de Ingeniería Agronómica, Campus de Excelencia Internacional Regional "Campus Mare Nostrum", Instituto de Biotecnología Vegetal, Escuela Técnica Superior de Ingeniería Agronómica, Universidad Politécnica de Cartagena, Cartagena, Spain*

KONSTANTINOS P. KOUTSOUMANIS • *Department of Food Science and Technology, School of Agriculture, Faculty of Agriculture, Forestry and Natural Environment, Aristotle University of Thessaloniki, Thessaloniki, Greece*

João Borges Laurindo • *Federal University of Santa Catarina, Department of Chemical and Food Engineering, Center of Technology, Florianópolis, SC, Brazil*
Alexandra Lianou • *Division of Genetics, Cell and Developmental Biology, Department of Biology, University of Patras, Patras, Greece*
Daniel Angelo Longhi • *Federal University of Paraná, School of Food Engineering, Jandaia do Sul, PR, Brazil*
Wiaslan Figueiredo Martins • *Food Technology Department, Goiano Federal Institute of Education, Science and Technology, Goiás, Brazil*
Peter Myintzaw • *Department of Biological Sciences, Munster Technological University, Cork, Ireland*
Thomas P. Oscar • *U.S. Department of Agriculture, Agricultural Research Service, Chemical Residue and Predictive Microbiology Research Unit, Center for Food Science and Technology, University of Maryland Eastern Shore, Princess Anne, MD, USA*
Wilmer Edgard Luera Pena • *Department of Food Technology, Federal University of Viçosa, Viçosa, Minas Gerais, Brazil*
Fernando Pérez-Rodríguez • *Department of Food Science and Technology, UIC Zoonosis y Enfermedades Emergentes ENZOEM, ceiA3, Universidad de Córdoba, 14014, Córdoba, Spain*
Maísa Daré Perim • *Department of Food, Faculty of Pharmacy, Federal University of Minas Gerais, Belo Horizonte, MG, Brazil*
Çağla Pinarli • *Department of Nutrition and Dietetics, Istanbul Gedik University, Istanbul, Turkey*
Arícia Possas • *Department of Food Science and Technology, University of Córdoba, Córdoba, Spain*
Angie Dahiana Duque Rodriguez • *Department of Food Technology, Federal University of Viçosa, Viçosa, Minas Gerais, Brazil*
Andrea Serraino • *Department of Veterinary Medical Sciences, University of Bologna, Ozzano Emilia, BO, Italy*
Fatih Tarlak • *Department of Nutrition and Dietetics, Istanbul Gedik University, Istanbul, Turkey*
Federico Tomasello • *Department of Veterinary Medical Sciences, University of Bologna, Ozzano Emilia, BO, Italy*
Antonio Valero • *Department of Food Science and Technology, University of Córdoba, Córdoba, Spain*

Chapter 1

Basic Concepts for Predictive Microbiology

Alice Karine da Silva, Maísa Daré Perim, Luma Moura Brito, and Verônica Ortiz Alvarenga

Abstract

Advances in science and technology have led to the generation of data that require storage, processing, and interpretation. Predictive microbiology is a valuable tool that merges microbiology, mathematics, and statistics to improve food safety and quality from microbial growth, survival, or inactivation parameters. Predictive models seek to understand the effect of environmental conditions (pH, temperature, aw, etc.) on microbial responses by mathematical models. The predictive models are developed by laboratory tests combined with software to predict the microbial kinetic parameters under different conditions. Indeed, the models do not replace microbial controls but are an additional device for decision-making supported by data-driven. Thus, this chapter aims to provide an overview of predictive microbiology, covering fundamental concepts, methodologies, types of models, and applications. Moreover, this chapter highlights basic concepts that need to be considered while performing predictive modeling and the limitations of this tool.

Key words Primary modeling, Secondary modeling experimental design, Food safety

1 Introduction

In recent decades, food quality and safety concerns have been growing. There are physical, chemical, and biological risks, which can affect food integrity and safety [1]. Foods are subject to food contamination by microorganisms, pathogenic or spoilage bacteria, which can cause food batch recall and foodborne diseases and impact the consumer's safety [1]. Due to the expansion of the food trade, it is possible to notice a relative difficulty in managing risks and ensuring the protection of consumers' health [2].

The unit operations in the food chain can affect microorganism viability leading to inactivation. Thereby, it becomes crucial to know the microbial behavior parameters to understand microbial growth dynamics [3] and the responses of microorganisms to specific environmental conditions [3, 4]. Data collection in different environmental conditions enables assessing microbial kinetics and can predict responses in other similar environments through

Verônica Ortiz Alvarenga (ed.), *Basic Protocols in Predictive Food Microbiology*, Methods and Protocols in Food Science, https://doi.org/10.1007/978-1-0716-3413-4_1,

mathematical models [5]. Predictive microbiology uses mathematical models to quantify the effect of intrinsic and extrinsic factors (i.e., temperature, pH, water activity, autochthonous microbiota, and natural antimicrobial compounds) on microbial behavior. Predictive models can predict growth, inactivation parameters, and toxin production [6, 7]. The responses provided by the models can support food processors and regulatory agencies in data-driven decision-making.

The first description using a model was done by Bigelow and Esty [8], Bigelow [9], and Esty and Meyer [10]. Although the concept of "predictive microbiology" was first proposed in 1937, it was not thoroughly applied until the early 1980s, when the response to large food poisoning outbreaks spurred efforts to apply mathematical models. These efforts were used in pathogen inactivation (for instance, for *Clostridium botulinum* and *Staphylococcus aureus*) and measuring the spoilage bacteria growth [11]. In the 1960s and 1970s, studies applied mathematical models aimed toward the inactivation of bacteria and fungi. The predictive microbiology field was raised in the 1980s and 1990s. This intensification was attributed to accessibility to computational tools and software, which allowed the use of more complex and more accurate models [12].

Predictive microbiology aims to mathematically represent a microbiological process's reality and quantify its intrinsic and extrinsic effects [13]. Predictive microbiology has emerged at the interface of different areas of knowledge, including microbiology, statistics, mathematics, and computation. It has become an essential tool for data-driven decision-making [14]. Due to the importance of individual factors for each food type, data collection for predictive microbiological models is mainly based on laboratory data. However, there is also an increasing amount of research focused on purposely contaminated food to generate data for predictive microbiology [6].

The US Department of Agriculture-Agricultural Research Service (USDA-ARC) and the UK Ministry of Agriculture, Fisheries and Food proposed the first predictive microbiology software's approach in the 1990s [15]. The systems were called Pathogen Modeling Program and Food Micromodel, respectively. Both tools have a database and mathematical models to describe growth responses to environmental factors of foodborne pathogens [4]. Afterward, other tools emerged for predictive microbiology, such as Seafood Spoilage Predictor (currently called Food Safety Spoilage Predictor) [16], Dmfit, Ginafit [17], Microbial Responses Viewer [18], MicroHibro [19], Combase [20], and Bioinactivation [21].

The guarantee of food safety and quality can be impacted by the emergence of innovative food products such as new preserving food technologies. Thus, investigations that evaluate possible complications that may affect the quality of products are necessary. Also, current knowledge that are already available, the effects of processing new products, and other factors are of great importance to consider when developing predictive models [14].

Additionally in the food industry, authorities are seeking solutions and tools to mitigate or solve problems related to food safety. Considering the challenges to maintain the food quality and safety throughout the food chain, it is feasible to apply predictive modeling in the food industry [11]. Therefore, it is essential to plan the entire process and collect microbiological data that adequately reproduce the behavior of the microorganism in that specific study environment as well as choosing the appropriate model which relies the studies [22].

This book will cover the particularities of food matrices, the relationship between foods and microorganisms, and data collection for predictive models. Therefore, understanding microbial behavior and the influence of environmental conditions is fundamental for properly developing studies based on predictive food modeling.

2 Application of Predictive Models

The application of predictive models serves as a science-based tool that can aid shelf life studies and help design or reformulate food products based on a safety and quality perspective, meeting ongoing needs for quality and food safety [23]. The predictive microbiology applications are shown in Fig. 1.

Predictive models provide a quantitative view of the effect of intrinsic (*see* **Note 1**) and extrinsic factors (*see* **Note 2**) on the quality and safety of food [14]. For instance, in quantitative microbial risk assessment (QMRA), predictive models can estimate the impact of unit operation along the food chain on microbial behavior and quantify the risks associated with contaminated food consumption [24]. Furthermore, the predictions by the models can support decision-making processes, highlighting its integration into self-control systems, such as hazard analysis and critical control points (HACCP), to determine process criteria and control limits [25]. Predictive microbiology can also optimize and validate new thermal and nonthermal food processing technologies due to its ability to assess the process impact against microbial inactivation and describe the response of microorganisms if there is any environmental change [26]. Therefore, predictive microbiology is a valuable tool for food quality and safety decision-making.

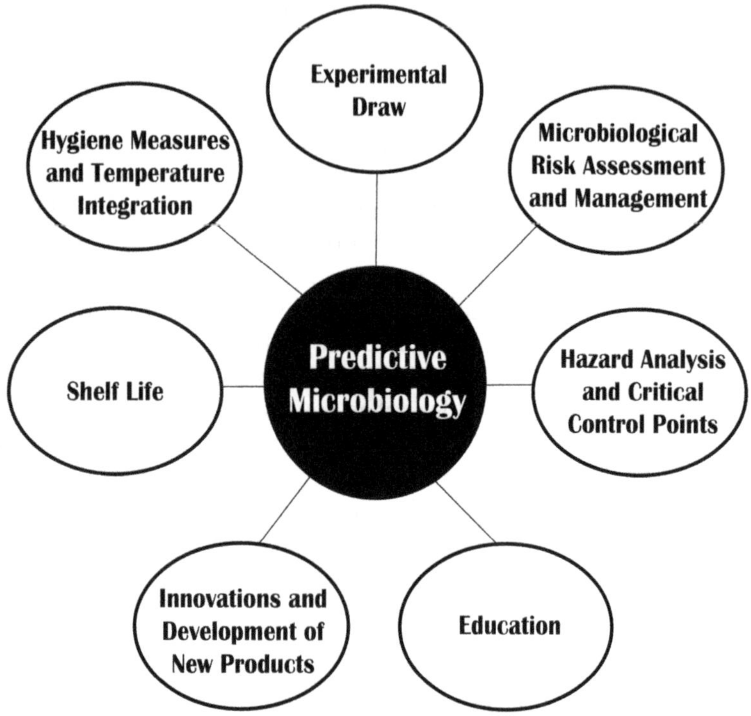

Fig. 1 Application of predictive microbiology

3 Concepts for Predictive Microbiology and Mathematical Modeling

An appropriate understanding of mathematical and microbiological concepts is essential for developing sound foundations for predictive microbiology. This understanding is necessary for creating realistic models and exploring various mathematical tools and methods in this field. For instance, to determine whether a proposed inactivation technology can be used, it is crucial to have an essential understanding of the characteristics of microorganisms and other relevant concepts in microbiology.

3.1 Microbiological Concepts

The logarithmic scale is commonly used to represent microbial growth and express populations graphically. It allows for better visualization and understanding of growth stages on a curve. Within an analytical scenario, various techniques serve to enumerate microorganism purposes and can be divided into two major groups: quantitative and qualitative analysis methods. Quantitative analysis involves numerically measuring a compound, group of compounds, or parameters (*see* **Note 3**), such as calculating colony forming units (CFU) per gram or milliliter, most probable number (MPN) [27] (*see* Chapter 3), or number of cells by microscopy [28] (*see* Chapter 7). In contrast, the qualitative analysis does not measure the specific amount of a compound or the value of an analysis parameter. Instead, it aims to indicate the presence or absence of

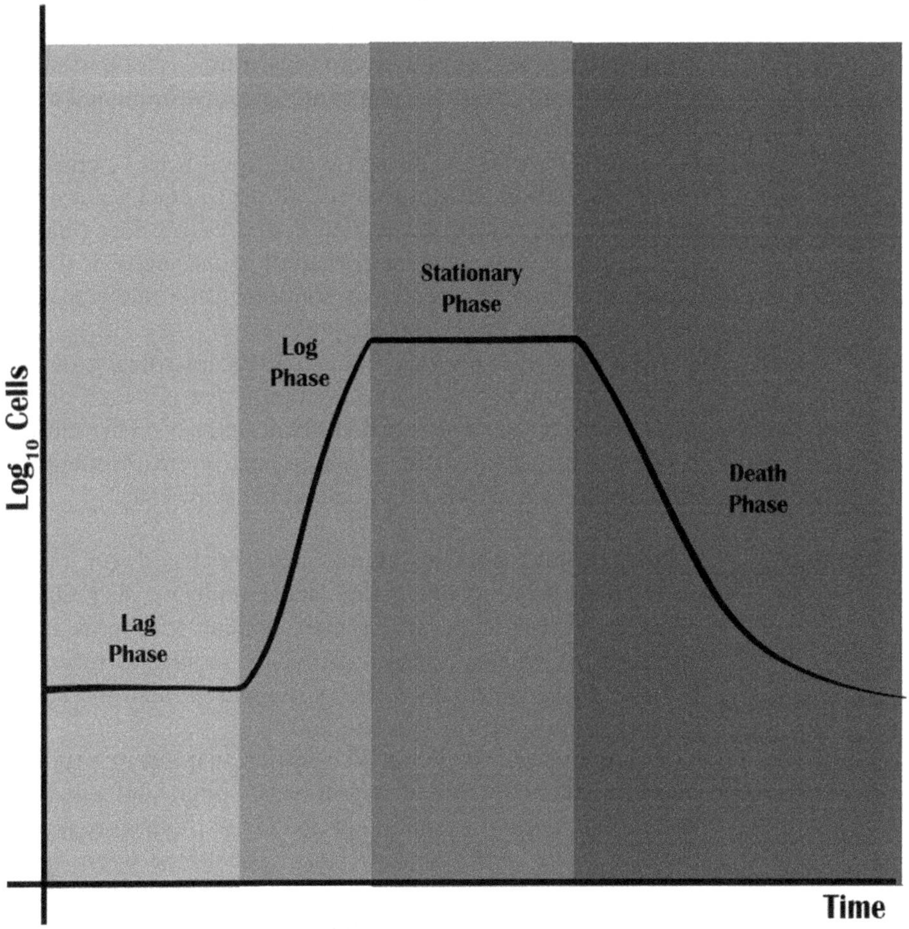

Fig. 2 Microbial growth curve

the analyzed parameter or boundaries for microbial growth (*see* Chapter 6).

Microbial growth involves increasing the number of individuals (cells) through processes such as binary fission, budding, spore formation, or fragmentation. This growth is characterized by a "generation time," which is required for a cell to divide (*see* Chapter 7) (assuming binary fission as the most common process) or for a population to duplicate itself. The generation time may vary depending on the microorganism or medium temperature [29]. Understanding the relationship between generation time and microbial growth phases is crucial, including the comparison between phases [30]. Microbial growth occurs in four stages, as Peleg and Corradini [31] showed in Fig. 2.

Lag phase: This phase can last for an hour or even several days, during which cells are in a state of latency growth and adjusting the metabolism before starting exponential growth. There is little or no cell division during this phase, resulting in no significant increase in the number of cells.

Log phase or exponential growth phase: Following the end of the latency state, a phase of increased metabolic activity begins. During this phase, cells undergo logarithmic cell division, and the generation time becomes constant. The environmental conditions influence this phase.

Stationary phase: This phase corresponds to a period of equilibrium in which the number of cell death equals the number of new cells, and metabolic activity decreases. Decline phase or cell death: In this phase, the number of dead cells exceeds the number of new cells, and this trend may continue until only a small fraction of cells remains, or no cells are present.

It is important to observe the different microbial growth phases when analyzing processes for developing new products or assessing the effectiveness of microbial control methods. This can help use predictive models to estimate the impact of environmental factors and process conditions on microbial behavior [13].

3.2 Mathematical Concepts

Predictive microbiology approaches are based on operational research applications and mathematical modeling [32]. Operational research term, also known as management science, is a scientific approach to solving complex problems, aiming to make the best possible decision, usually in a scenario with limitations, such as the scarcity of resources in a system [33].

Mathematical modeling is an operational research tool to elaborate the decision-making system or to understand a complex and real problem logically and integrally. These models are represented mathematically by equations [34]. Usually, it is an empirically elaborated model (Fig. 3) based on laboratory result observation in a controlled environment, and it aims to predict the behavior of a system to guide decision-making about a problem situation.

For this, it is necessary to understand deeply the variables (*see* **Note 4**) involved in the problem and the nonmathematical issues surrounding it, for example, the reasons for the limitations imposed

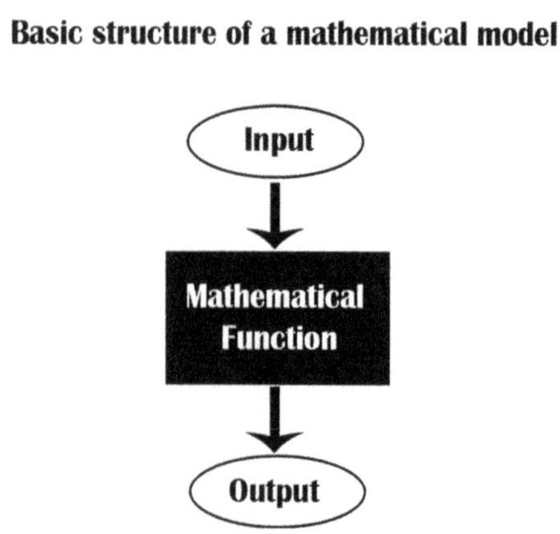

Fig. 3 Schematic representation for model development

on the model, either because of the search for low-cost production or the impossibility of having a variable with a negative value. Then, it must analyze which resources of reality will be used in constructing the model and which ones should be ignored, aiming at optimizing the system, with a good fit (or with low bias) and its accuracy compared to reality [35]. Modeling helps us understand biological systems better and can help us make predictions about their behavior. A model refers to a mathematical equation or set of equations that represent a biological process, system, or relationship. Such models usually involve several random variables and a combination of variables, constants, or parameters. A model that doesn't include random variables is referred to as deterministic.

Different types of modeling are available, such as descriptive, prescriptive, optimized, static or dynamic, linear or nonlinear, complete or incomplete, deterministic or stochastic, and network models, each of which can be suitable for different problems and scenarios [34]. A predictive model aims to describe the behavior of a natural phenomenon, raising its variables and how they relate to the process in a deterministic way [13]. Some examples of mathematical models within the microbiological context are inactivation curves, growth of a given microorganism, and heat penetration curves, among others [36]. Table 1 shows the different types of

Table 1
Description of terminology used for model in predictive microbiology

Model	Description
Static model	A static mathematical model is designed to perform decision-making in a fixed period, where there is no sequence of decisions to be taken in the long term
Dynamic model	This model is used when the objective is to find decisions over some time. Thus, a series of decisions are to be taken in this interval, where the variables can be changed over time
Linear model	Based on the multiplication and addition of constants to the decision variables of the objective function and its restrictions
Nonlinear model	A nonlinear model occurs when an optimization model is not linear, which usually involves a higher degree of complexity, and can form a convex, concave function, non-convex nor concave function, or a convex and concave function
Deterministic model	One that has a known number of inputs and outputs. In this way, it does not present random variables as it represents the evolution of the system in a certain period of time
Stochastic model	The stochastic model, as well as the deterministic one, evaluates the evolution of the system in a certain period of time. However, this model, unlike the previous one, presents random variables. Due to uncertain variables of the system, the measurement of time in the system can be continuous or in intervals
Network models	Network models are a way of optimizing mathematical models using programming techniques or graphical techniques

Adapted from Winston and Goldberg [34]

mathematical models that can be applied in predictive microbiology.

Linear mathematical models are characterized by presenting a linear function, called an objective function, with independent variables ($x1;x2;xn$ ...) related to linear constants (an. xn) that can be both equalities (=) and inequalities (≥ or ≤). Constraint elements are also involved in this system, applied under the conditions of the decision variables. For example, they are determining whether a variable can or cannot have a negative value. Meanwhile, nonlinear models can form a convex, concave function, and others. Also included in this group of mathematical models are functions of zero order, first order, second order, and others, leading to different ways of optimizing and solving the model [37].

System resolution can use various techniques, including simplifying the model, dividing it into subproblems, and simplifying the equations without considering the system's restrictions. Another critical point inside nonlinear models is the polynomial model. A group is also known as the response surface model (Fig. 4) [38]. The characteristic is that its objective function and constraints are well-defined functions, nonlinear in polynomial format, which does not always occur in all nonlinear mathematical models. One of the defining features of this type of model is that its objective function and constraints are well-defined, nonlinear polynomial functions. This is not always the case with other nonlinear mathematical models. These models can handle problems involving sine, radicals, and logarithmic functions, making them useful in predictive microbiology, where multiple determinations must be made simultaneously [39]. They are classified as secondary and empirical models, often called "black box" models. By using terms of the first, second, third, and fourth-order (quadratic function), it is possible to simultaneously evaluate the effect of multiple variables on the system being researched in the polynomial form [40–42]. However, it's important to note that the function may demonstrate exponential growth as the number of variables increases.

Another type of model is the stochastic model, which can be used in consumer brand choice research or microorganism growth studies, taking into account multiple random behavior variables, such as temperature, pH, and storage temperature variability. In addition, it seeks to describe the uncertainty and variability of the microbial response in the system [43].

Finally, network models can form artificial neural networks, representing the interdependence between the factors involved in the system. They are a type of artificial intelligence that has been emerging as a promising tool in treating biological data, for example, developing models to describe microbial growth in specific culture media. After all, artificial neural networks (ANN) can identify multiple parameters of a model (linear or nonlinear) in a

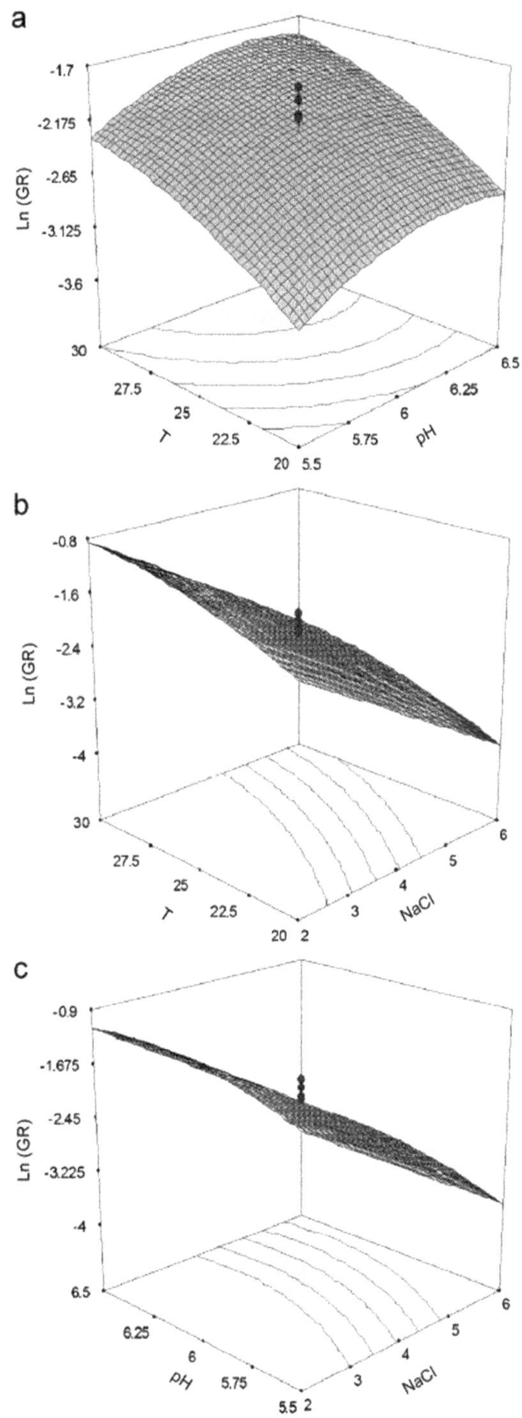

Fig. 4 The 3D surface plots of growth rate (GR) of *Clostridium sporogenes* spores affected by (**a**) temperature (T) and pH, (**b**) temperature (T) and NaCl, and (**c**) pH and NaCl of RS models. (Reuse with permission. Dong et al. [38])

discriminative way, organizing them logically for analysis, even coming directly from laboratory results (black box models), which makes decision-making of the problem situation more complete and increasingly data-driven, along with more minor estimated errors [44].

These networks can perform parallel calculations of several functions of the developed model. For this to happen, it is necessary to draw the neural layers and the mathematical tasks between them (whether linear or nonlinear, simple or polynomial) and determine the level of connectivity between one vertex and another [45]. In order to fully comprehend predictive modeling, mathematical and statistical concepts are essential. These concepts encompass variables, optimal and suboptimal solutions, objective function, model constraints, sensitivity analysis, regression analysis, prediction, parameters, and simulation.

Different mathematical models can describe and predict microbial behavior in food systems in predictive microbiology. White box or mechanistic models are based on the underlying mechanisms and physical processes that govern microbial growth and survival, such as nutrient uptake, metabolism, and environmental conditions. These models are typically more complex and require more detailed information about the system. Still, they can provide a deeper understanding of the underlying biological processes and be used to design more effective control strategies. On the other hand, black box models are generated purely from experimental data and do not necessarily rely on a deep understanding of the underlying mechanisms [47]. These models are often more straightforward but may not provide as much insight into the underlying biological processes.

Finally, some intermediate models combine both white box and black box approaches. These models may incorporate some mechanistic knowledge to improve the accuracy and robustness of the model while still relying on experimental data for parameter estimation and validation. Overall, the choice of model type depends on the specific research question, available data, and level of understanding of the system [14].

3.3 Types of Predictive Models

Predictive models are becoming valuable and fast tools in the search for answers to specific problems and can be used for several purposes, i.e., models of growth (*see* Chapter 4) and microbial death (*see* Chapter 5) as well as boundary conditions such as growth versus no-growth (*see* Chapter 6). Predictive models can be utilized to evaluate microbial growth, survival, and inactivation, as well as boundary conditions such as growth versus no-growth. These models can be classified through primary, secondary, and tertiary levels [48].

3.3.1 Primary Models

Primary models measure the microorganism response over time for a single set of conditions, allowing growth and decline curves to be generated. The responses provided by primary models are inactivation and growth rate, delay time, or times of turbidity/toxin formation [49].

3.3.1.1 Growth Models

Primary growth models are divided into sigmoidal and mechanistic functions. Sigmoidal curves describe the effect of biochemical reactions on microbial growth rate [50]. There are two approaches for primary growth models: sigmoidal and mechanistic functions. Sigmoidal curves are used to depict the impact of biochemical reactions on microbial growth rate. The most commonly used models for this are the logistic (Eq. 1) [50] and Gompertz modified (Eq. 2) [51] models. On the other hand, mechanistic approaches reveal that the maximum specific growth rate and the lag depend on the environment. Baranyi and Roberts (Eq. 3) [52] came up with a mathematical model that demonstrates an inversely proportional lag to the maximum specific growth rate:

$$\Upsilon(t) = \Upsilon_0 + \Upsilon_{\max} - \ln\{\exp(\Upsilon_0) + [\exp(\Upsilon_{\max}) - \exp(\Upsilon_0)]\exp(-\mu_{\max} t)\} \quad (1)$$

$$\Upsilon(t) = \Upsilon_0 + (\Upsilon_{\max} - \Upsilon_0)\exp\left\{-\exp\left[\frac{2.71\mu_{\max}(t_{\text{lag}} - t)}{\Upsilon_{\max} - \Upsilon_0} + 1\right]\right\} \quad (2)$$

$$\begin{cases} \Upsilon(t) = \Upsilon_0 + \mu_{\max} A(t) - \ln\left[1 + \dfrac{\exp(\mu_{\max} A(t) - 1}{\exp(\Upsilon_{\max} - \Upsilon_0)}\right] \\ A(t) = t + \dfrac{1}{\mu_{\max}} \ln\left[\exp(-\mu_{\max} t) + \exp(-\mu_{\max} t_{\text{lag}}) - \exp(-\mu_{\max} t - \mu_{\max} t_{\text{lag}})\right] \end{cases} \quad (3)$$

where $\Upsilon_{\max}$ is maximum population; Υ_0 is initial population; $\mu_{\max}$ is maximum growth rate; and t_{lag} is the lag time.

3.3.1.2 Inactivation Models

Predictive models can be used to describe microbial kinetics inactivation resulting from thermal or nonthermal processes. Microbial populations are typically measured at discrete time points to construct an inactivation curve. The resulting survival curve may show linear or nonlinear behavior.

Primary inactivation models allow us to estimate inactivation parameters such as inactivation rate ($k_{\max}$), shoulder length (SI), and tail formation (residual population) (y_{res}). Regarding describing inactivation behavior, several authors developed equations for this purpose. The linear model [8] (Eq. 4) was the first inactivation model to apply fit inactivation kinetics. Nonlinear models with shoulder and tail [53] (Eq. 5) describe a lag time before the inactivation and a residual population after the treatment, tail region. Further, the concave and convex curves, due to biological variations in inactivation, may be expressed as a statistical model of the distribution of inactivation times, Weibull model [41] (Eq. 6):

$$y(t) = y_0 - \frac{t}{D} \tag{4}$$

where y_0 is initial population and D is decimal reduction time.

$$y(t) = y_{\mathrm{res}} + \log\left[\frac{(10^{y_0 - y_{\mathrm{res}}} - 1)\ \exp(k_{\max} S_l)}{\exp(k_{\max} t) + \exp(k_{\max} S_l) - 1} + 1\right] \tag{5}$$

where y_{res} is residual population, y_0 is initial population, $k_{\max}$ is inactivation rate, and S_l is shoulder length.

$$y(t) = y_0 - \left(\frac{t}{\delta}\right)^p \tag{6}$$

where y_0 is initial population, δ is time for first decimal reduction, and p is curvature parameter.

3.3.2 Secondary Models

Secondary models describe the effects of multiple variables on microbial behavior in food. These models are based on the primary models, which describe microbial growth, survival, or inactivation as a function of environmental factors such as temperature, pH, and water activity. Secondary models consider additional factors influencing microbial behavior, such as other microorganisms, food components, and chemical or physical treatments. Examples of secondary models include competitive growth models, which describe the growth of multiple microorganisms in a food matrix, and hurdle models, which represent the effect of numerous preservation factors on microbial growth or survival. Other secondary models include models for the impact of preservatives, thermal processing, and modified atmosphere packaging on microbial growth. These models can approach kinetics or probability description for the influence of intrinsic and extrinsic factors on either microbial growth, survival, or inactivation [49]. Probability models are developed through regression analysis to estimate the effects and interactions of independent variables, allowing for the assessment of the probability of toxin production, spore germination, or microbial growth boundaries. Meanwhile, kinetic models use mathematical functions, such as linear or nonlinear regressions and polynomial regressions, to assess the impact of environmental conditions on microbial kinetic parameters, i.e., inactivation or growth rate, lag phase duration, and maximum population density. Polynomial functions (Eq. 7) are widely employed in predictive microbiology to model the effects of environmental factors simultaneously [54]. This approach is commonly utilized to develop the increasingly popular square root [55] (Eq. 8) and cardinal parameter-type models [56] (Eq. 9). These models individually consider each environmental factor and then create a general model that describes their combined effects:

$$y = \beta_0 + \sum_{j=1}^{k} \beta_j X_j + \sum_{j=1}^{k} \beta_{jj} X_j^2 + \sum_{j \neq 1}^{k} \beta_{jl} X_j X_l + \varepsilon \qquad (7)$$

where β_0, β_j, β_{jj}, β_{jl} are the estimated coefficient regression, X_j and X_l are independent variables, and ε is error.

$$\sqrt{\mu_{\max}} = b\,(T - T_{\min}) \qquad (8)$$

where $T_{\min}$ is the minimum temperature below which the maximum growth rate is equal to 0 and obtained through a linear regression of the square root of the maximum growth rate temperature.

$$\mathrm{CM}_n(X) = \begin{cases} 0, & X \leq X_{\min} \\ \dfrac{(X - X_{\max})(X - X_{\min})}{(X_{\mathrm{opt}} - X_{\min})^{n-1}\left[(X_{\mathrm{opt}} - X_{\min})(X - X_{\mathrm{opt}}) - (X_{\mathrm{opt}} - X_{\max})((n-1)X_{\mathrm{opt}} + X_{\min} - nX)\right]}, & X_{\min} < X < X_{\max} \\ 0 & X \geq X_{\max} \end{cases} \qquad (9)$$

where X is temperature, pH, or aw; $X_{\min}$ and $X_{\max}$ are values of X below and above which no-growth occurs, respectively; X_{opt} is the value X at which microbial growth is optimum; and n is a shape parameter. In optimal condition de $\mathrm{CM}_n(X_{\mathrm{opt}})$, the value is equal 1 and for $X_{\min}$ and $X_{\max}$ the CM_n is equal to 0 [57].

3.3.3 Tertiary Models

Finally, the tertiary level refers to computational tools that integrate primary and secondary models in friendly interfaces [58]. The tertiary level of predictive microbiology involves the integration of primary and secondary models into computational tools that provide user-friendly interfaces for predicting microbial responses. One of the primary challenges in predictive microbiology is managing large amounts of data and retrieving specific information efficiently. Decision-support tools, such as computer software programs, can help to address these challenges by providing users with a simplified visualization of model inputs and outputs through graphical interfaces. However, the development and application of information management systems in predictive microbiology have been limited compared to other scientific fields. This highlights the need for continued investment and development to ensure the predictive models are accurate and up-to-date. Overall, the tertiary level of predictive microbiology plays a critical role in advancing our understanding of microbial behavior and improving food safety. The development and integration of advanced computational tools may lead to play a critical role in improving our ability to predict and manage the behavior of microorganisms in a wide range of applications [59].

3.3.3.1 Types of Programing Languages, Tools, and Software Available to Predictive Microbiology

Tertiary models have become an important, if not indispensable, tool for elaborating and resolving complex problems related to assessing microbial behavior. Considering this context, software are becoming more common and enable more robust data analysis by considering a greater number of variables. It is also possible to use more comprehensive tools for modeling, such as Microsoft Excel and programming languages like VBA, R, and Python. However, providing a user-friendly interface for estimating microbial parameters is more beneficial for food processors, regulators, and the general public [14]. Regarding food safety management and assistance in decision-making, tertiary models provide insights into achieving food safety management measures in a short timeframe. Thus, different databases, software, and add-ins have been developed for predictive microbiology in the last 20 years. Table 2 presents a survey of characteristics and applications of some devices used in predictive microbiology.

4 How to Design to Predictive Microbiology: Collect Data and Design Experiment

Developing a predictive model requires careful planning and execution of the experimental design for evaluating the effect of intrinsic and extrinsic factors. A good experimental design ensures high-quality and reliable results from the data gathering. In addition, adequate planning and the organization of the experiment are crucial for ensuring the collected data has the necessary quality and microbial behavior to be studied [75].

During planning stage, the model should be designed to consider several parameters such as the central hypothesis of the study, the food matrix and its particularities, the target microorganism behavior in this matrix, the adequate volume of data to support the model, and the necessary resources [76].

Environmental factors that affect microbial behavior and how their interaction occurs should also be considered during the design process. Thus, it may lead to increase the chances of extracting the desired information and reduce the probability of excessive experimental work due to failure at the end of the process [77].

The experimental design should control and adapt the factors according to the design objectives and the hypothesis it aims to test. The physiological state of inoculum cells, substrate or culture medium needed, dependent variables proposed for the model, combinations of factors that may be included in the model, and how collected data are carried out are all important factors to consider during the designing process [76, 78].

The factors that must be controlled must be planned and adapted according to the experimental design objectives and the hypothesis being tested. Within the design process, it is essential to

Table 2
Tertiary models for predictive microbiology

Software/tool	Characteristics and applications	Available in	References
Biogrowth	Biogrowth is another R package used in predictive microbiology. Used to build primary and secondary models under statistical and dynamic conditions based on growth data to generate growth curves. It is possible to make predictions with it under isothermal and dynamic conditions and generate predictive intervals	https://cran.r-project.org/web/packages/biogrowth/vignettes/v02_growth_fitting.html	[60]
Bioinactivation	The Bioinactivation is available as an R package, as a Shiny app, and allows the building of inactivation curves under different conditions. For example, it is possible to create an inactivation curve under uncertain parameters using stochastic models or model predictions using dynamic conditions. The Shiny app allows the programmer to import data from Microsoft Excel too	https://foodlab-upct.shinyapps.io/bioinactivationFE/	[21]
BioOED	BioOED is a package for R programming language specific to predictive microbiology. It is used for inactivity characterization through sensitivity analysis and optimum experiment design	https://cran.r-project.org/web/packages/bioOED/vignettes/vignette_bioOED.html	[61]
Combase	Online database focused on describing microorganisms' survival and growth under various food matrices. Combase can be used to develop new food products, reformulate foods, design challenge test protocols, produce food safety plans, and help public health organizations develop science-based food policies through quantitative risk assessment. It is a friendly interface and may be accessed on any platform, including mobile devices	www.combase.cc	[20]
D-database	Database with 5000 records for microbial inactivation parameters. This platform provides information to develop inactivation models, including metaregression	https://foodmicrowur.shinyapps.io/Ddatabase/	[62]
DMFit	It is used to determine the microbial growth curves based on the Baranyi model (three versions), three-phase linear model, two-phase linear model (two versions), and the linear model. It is available as an online version inside Combase and as an Excel add-in offline version	http://www.combase.cc/index.php/en/tools	[20]

(continued)

Table 2 (continued)

Software/tool	Characteristics and applications	Available in	References
FDA-iRisk	Online system designed to analyze data related to microbial and chemical risks in food. To evaluate risk on this platform, it is necessary to include food consumption, processing, and preparation data. With this information, estimating the risk associated with food consumption is possible	https://irisk.foodrisk.org	[63]
FSK-Lab (Food Safety Knowledge Lab)	It is an open-source risk assessment extension derived from the data analytics platform, graphical programming framework, and Konstanz Information Miner (KNIME) of the Risk Assessment Modeling and Knowledge Integration Platform initiative. There, it is possible to edit, create, and execute models. The models created or edited on this extension are interchangeable with other software	https://foodrisklabs.bfr.bund.de/fsk-lab/	[64]
FoodRisk	FoodRisk is a metadatabase of tools and models that can generate information for risk analysis as it pertains to the safety of food. Can be used in predictive microbiology studies. This site is operated by Joint Institute for Food Safety and Applied Nutrition (JIFSAN) in collaboration with the Center for Food Safety and Applied Nutrition from the US Food and Drug Administration (CFSAN/FDA) and the Food Safety and Inspection Services from the US Department of Agriculture (FSIS/USDA)	https://www.foodrisk.org/	[63]
Food Spoilage and Safety Predictor (FSSP)	The Food Spoilage and Safety Predictor software contains various models to predict the effect of product characteristics and storage conditions on food's shelf life and safety. This software can simulate a cooling chain to estimate the number of days the supermarket can sell the product and the shelf life. Also, it can evaluate the impact of new formulation on shelf life product and assess the effect of temperature changes on food safety for different food matrices	http://fssp.food.dtu.dk	[65]

GinaFit	Excel add-in for inactivation curves with ten model types: (i) classical log-linear curves, (ii) curves displaying a so-called shoulder before a log-linear decrease is apparent, (iii) curves displaying a so-called tail after a log-linear decrease, (iv) survival curves displaying both shoulder and tailing behavior, (v) concave curves, (vi) convex curves, (vii) convex/concave curves followed by tailing, (viii) biphasic inactivation kinetics, (ix) biphasic inactivation kinetics preceded by a shoulder, and (x) curves with a double concave/convex shape	https://cit.kuleuven.be/biotec/software/GinaFit	[17]
Growth rates	Growth rates is an R package that brings methods to determine microbiological growth rates using experimental data. This package focuses particularly on batch experiments and plate reader trials	https://cran.r-project.org/web/packages/growthrates/vignettes/Introduction.html	[66]
IPMP Global Fit	This is a software tool for Windows that can be used to fit a variety of inactivation models to data using a one-step approach. However, it does not include features to fit data gathered under dynamic conditions or to fit models using the two-step approach	https://www.ars.usda.gov/northeast-area/wyndmoor-pa/eastern-regional-research-center/docs/ipmp-global-fit/	[67]
Listeria cardinal models	The device simulates *Listeria monocytogenes* behavior using meta-analytical data	https://vcadavez.shinyapps.io/ListeriaCardinalModel/	[68]
MicroHibro	It makes predictions from different food matrices/culture media or distinct microorganisms. It is software and data-based, divided into two modules. The first one is used to build predictions/simulations under specific conditions standard by the author. This is possible using mathematical models in their repository database originating from the scientific literature. The second module consists of an object-oriented system. MicroHibro can be used to establish and build models for shelf life, risk assessment, sampling plans, and microbial prediction models. Besides, it can be used for the validation of microbiological models	https://microhibro.com/en/assessment	[19]

(continued)

Table 2 (continued)

Software/tool	Characteristics and applications	Available in	References
MicroFit	MicroFit is a free, autonomous software with an objective (intuitive) interface that allows the development, adjustment, and comparison of mathematical models of microbial growth. Based on data analysis of microbial behavior, it is possible to establish the useful life of a product and make decisions based on the data presented by the software. It is independent of other tools	http://www.univasf.edu.br/~cpgea/arquivos/	[69]
Microbial Responses Viewer (MRV)	The MRV (Microbial Responses Viewer) offers details on growth and no-growth limits as well as the particular growth rates of selected microorganisms. This tool enables the retrieval of data related to bacterial growth and no-growth, as determined by specific temperature, pH, and aw environmental factors. Furthermore, MRV facilitates the retrieval of growth rate data for microorganisms under specific environmental conditions	http://mrviewer.info/	[18]
MicroRisk Lab	MicroRisk Lab is an R programming language with a friendly interface called Shinny. This tool aims to fit models to illustrate microbiological growth, inactivation, and interaction. This online freeware has proper functions for primary and secondary models. Likewise, it provides deterministic and stochastic simulation for isothermal or non-isothermal models	https://www.microrisklab.top/english/	[70]
nlsMicrobio	R package that includes various microbial inactivation models. However, they can only be used to describe data under isothermal conditions. This package allows the programmer to create competition curves, growth curves, secondary growth models, and others	https://cran.r-project.org/web/packages/nlsMicrobio/nlsMicrobio.pdf	[71]
Pathogens in Food	Database, which contains a systematic review of 14 biological hazards in various food matrices. This database has 6000 records of food pathogens	https://vcadavez.shinyapps.io/MeatProducts/	[72]

Pathogen Modeling Program (PMP)	Developed by the US Department of Agriculture-Agricultural Research Service (USDA-ARS), this software which has an online and offline version is a free-access tool and provides predictive models for some determined variables. It is possible to generate growth, death, cooling, heat inactivation, survival, and transfer microbiological prediction models for several stairs. Despite all these facts, it is important to point out the importance of validating or modeling for our specific culture media/food analysis	https://portal.errc.ars.usda.gov/PMP.aspx	[15]
Praedicere Possumus	It is a free-based software that can predict pathogenic microorganisms' growth and density using several environmental conditions. The results can be exported in CSV, MS Excel, or PDF formats	http://praedicere.uniud.it/	[73]
Sym'previus	This software tool online that can predict growth, growth/no-growth boundaries, and thermal inactivation of microorganisms using several environmental conditions. It helps to guarantee food safety and quality because it is designed to support shelf life determination, product formulation, and process optimization. Sym'previus is available in both a free and a premium version (for simulation tools)	https://symprevius.eu/en/	[74]

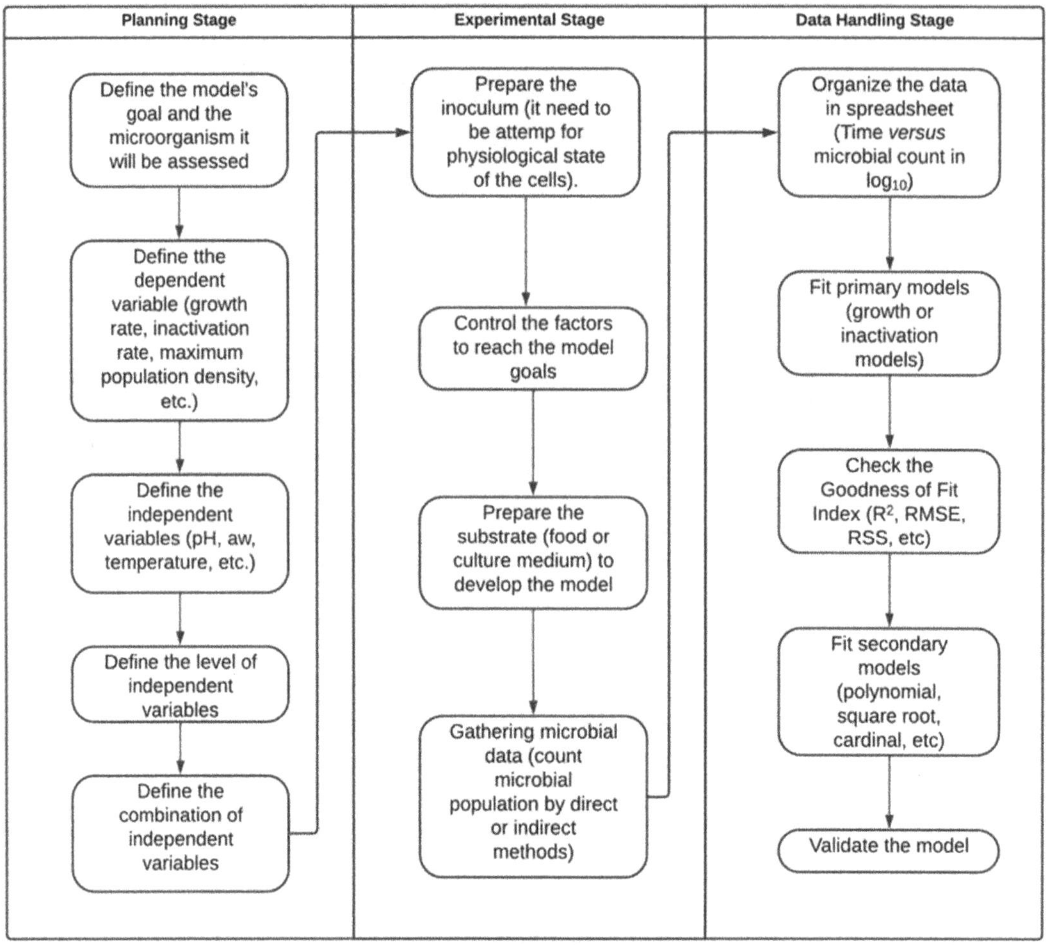

Fig. 5 Stepwise to develop a predictive model

consider other specific details such as the physiological state of inoculum cells, the substrate or coculture medium required, the dependent variables proposed for the model, combinations of factors to be included in the model, and the method of data collection. The factors to be considered are shown in Fig. 5 [77].

For the experimental design, a procedure containing two stages can be applied, as suggested by [14] (i) performing screening experiments in a wide range of factors and (ii) conducting a data collection study within the region of interest, including additional levels of factors that result in a more refined model and more accurate forecasts. Thus, experimental designs are used in predictive modeling to "reduce the number of experiments," making it an important stage for the development of models.

The collected data and the number of points needed to represent microbial behavior are crucial to reducing the uncertainties related to this behavior. Thus, the distribution of the collected points within the experimental design becomes fundamental for estimating the dependent variables. As a result, the representativeness of the model increases, and there is a reduction in the variance of the estimated parameters.

Experimental designs are used to model microbial responses in foods. The advantages of using practical methods correspond to the "ease of implementation together with its data processing." In addition, all combinations are explored, as all information will be obtained from the experiment, making it an easy tool to handle statistically. It is possible to develop the experimental design through complete factorial designs, fractional factorial designs, or central composite designs, depending on the number of variables in the experiment [77].

4.1 Design Experiment

4.1.1 Complete Factorial Design

The complete factorial design allows a complete investigation of all combinations of different variables, enabling direct modeling of interactions, in which it becomes possible to evaluate the environmental factors that influence the growth and inactivation of microorganisms. The complete factorial design has significant advantages, as it is easy to implement and process your data, as all combinations are explored. Nevertheless, this design has the disadvantage that a new factor/level is added to the experiment. There will be a significant increase in the number of experiments, which can become laborious and expensive [79]. The number of experiments that will be performed can be calculated as 2^k, where k corresponds to the number of independent variables in the experiment [80].

4.1.2 Fractional Factorial Design

Fractional factorial designs are commonly used in situations where there are a large number of independent variables, as they help to reduce the number of experiments required. These designs are based on prior knowledge or assumptions about the most important factors or expected interactions. However, they can be more complex to develop than complete factorial designs, as the different parameters that should be included in the experimental plan must be carefully considered [81].

In order to define the matrix of a combination of levels of independent variables, statistical software is often necessary. One type of fractional factorial experiment is the Box-Behnken design, which combines two-level factorial experiments with balanced incomplete blocks and includes additional experiments in the central points of the design to test the repeatability of the adopted model design. Another type of fractional factorial experiment is the factorial Latin square design, which corresponds to a Latin square of order x, an arrangement of x letters in an x by x matrix. In this design, each letter appears once in each line and column inside a square block or field [81, 82].

For experiments in predictive microbiology, the Latin square design typically corresponds to a primary factor known as "treatment," represented by randomly distributed letters appearing only once in a row or column. This design is useful when several blocking factors, called "annoyances," need to be controlled, even though the approach is not limited to the main factor of interest. The Latin square design corresponds to an example of an incomplete block. It can be extended to more individual factors, used as the Greco-Latin square or the Hyper Graeco-Latin designs [82, 83].

4.1.3 Central Composite Design

The central composite design is commonly used for predictive food modeling experiments and involves a complete factorial experiment with a limited number of levels per environmental factor, making it suitable for studying simpler model structures. This design includes a complete factorial experiment with a set of central points and two axial points on the axis of each design variable at a distance of a from the design center, as described by Eq. 10. The number of experiments for k variables can be determined using the equation, with n_0 representing the number of experiments at the central point. Additionally, a set of "star points" is included to estimate curvature and enhance the central points [84]:

$$2^k + 2k + n_0, (n_0 \geq 1) \tag{10}$$

where k is the number of independent variables and n_0 is the number of experiments of central point.

4.1.4 Doehlert Matrix

The Doehlert matrix is a type of experimental design that is characterized by points that are uniformly spaced on concentric spherical shells. This design is also known as a uniform shell design, and it aims to fill the experimental space uniformly [85]. The number of designs is determined by the number of variables, k, and the number of center points, n_0, according to the formula $k_2 + k + n_0$. For instance, if $n_0 = 1$ and $k = 4$, there will be 21 experiments. The center experiment can be repeated multiple times to estimate the observed variance, which can be used for model validation. One of the advantages of the Doehlert matrix is that it can be easily expanded by adding new variables or increasing the range of the tested parameters, providing flexibility for the experimental design [86].

4.2 Growth Matrices and Microbial Strain

Predictive modeling in microbiology requires an understanding of the unique characteristics of the matrix that provides conditions for microbial growth. The behavior of microorganisms can be explained through significant environmental factors such as temperature, pH, and water activity but also by structural compositions of the matrix. This fact is essential to consider in the development of predictive models. For example, microbial growth occurs more rapidly in a liquid medium than in food media [87]. Thus, when comparing predictions generated from liquid matrices with observations for food ones, it is necessary to consider the potential factors that may result in divergences. These factors may include the presence of native microbiota or additional environmental factors in food matrices that were not included in the models. For detailed information, *see* Chapter 3.

An appropriate microbial strain is critical in developing a reliable and accurate predictive model. Firstly, the chosen strain should be representative of the microbial population of interest in the food matrix and be relevant to the food matrix and its storage conditions. The selection encompasses the origin, including different strains of foodborne pathogens, physiology, and growth characteristics of the strain. Depending on the study purpose, strains or cocktails can be used. This allows the assessment of the variability related to resistance to stress, behaviors, and the influence of lineage variability given to changes in the environment between different lineages of the same species, thus increasing the representativeness of the situation found in the analyzed food [88]. Alternatively, surrogate microorganisms can be used instead of specific pathogens when the pathogen is not indicated. The surrogate should have similar behavior and kinetics characteristics, except for virulence, similar to the target pathogens [89].

The microorganisms, vegetative cells, or spores selected for the study must be incubated in standardized conditions, preferably similar to those found in food. Also, inoculum preparation is a crucial step for predictive model development. For more information about the selection and inoculum preparation and method of inoculation, refer to Chapter 2.

4.3 Goodness Index Fit and Model Validation

After elaborating and refining the mathematical model, it is necessary to test it to verify its accuracy in the face of reality, being the last step of the modeling cycle [90].

Validation methods can be divided into graphical and mathematical methods. The first one allows quick visualization and interpretation of the objective data and is characterized by creating a graph where the expected values according to the model and the values acquired in a laboratory way are plotted [91, 92]. Mathematical validation implies the calculation of statistical factors, for example, mean squared error, accuracy factors, and bias, among others [13, 93, 94].

After performing the curve fitting, the model must be validated using statistical indexes, constantly evaluating which indexes best apply to each type of model, taking into account its characteristics, such as whether the predictive model is a linear function or not [95]. For more information, *see* Chapter 10 for model validation.

4.3.1 Goodness Fit Index

The goodness fit indexes are used to evaluate the fit of a model. Some metrics, such as R^2, RSS, and RSME, can provide different information about the accuracy and precision of the model's predictions [96].

4.3.1.1 R-Squared

R-squared (R^2) is a statistical measure that indicates how well the model fits the data, that is, it evaluates the performance of a security or fund (dependent variable) in relation to a given reference index (independent variable) (Eq. 11). R^2 adjusted (Eq. 12) determines the extent of variance in the dependent variable that can be explained by the separate variable:

$$R^2 = \frac{\sum_{i=1}^{n}\left(\widehat{y}_i - \frac{1}{n}\sum_{i=1}^{n} y_i\right)^2}{\sum_{i=1}^{n}\left(y_i - \frac{1}{n}\sum_{i=1}^{n} y_i\right)^2} \tag{11}$$

$$R^2_{\text{adj}} = 1 - \left(1 - R^2\right)\frac{n-1}{n-k-1} \tag{12}$$

where n is the number of observations, y_i is the observed value for sample i, $\widehat{y}_i$ is the value of the prediction for sample i, k is the number of parameters, and n is the number of sample data.

4.3.1.2 RSS

Residual sum of squares (RSS), also referred to as the sum of squared errors (SSE), is a statistical method that indicates whether the regression model fits the actual dataset well or not. It measures the variance of the value of the observed data when compared to its value predicted by the regression model (Eq. 13):

$$\text{RSS} = \sum_{i=1}^{n} \left(y_i - \widehat{y}\right)^2 \tag{13}$$

where y_i is the observed variable value and $\widehat{y}$ is the value estimated by regression line.

4.3.1.3 RMSE

RMSE (root mean squared error) is a measure of absolute error that squares the deviations to prevent positive and negative deviations from canceling out. This measure also tends to overestimate large errors, which can help weed out methods with these errors (Eq. 14):

$$\text{RMSE} = \sqrt{\frac{\text{RSS}}{n-k}} \tag{14}$$

where n is the number of samples and k is the number of parameters.

5 Limitation

Although microbial predictive models have brought significant advancements to food safety, addressing some limitations associated with their development and application is essential. Adequate knowledge of the initial conditions, understanding the challenges encountered during the study of food matrices and production processes, and considering the variability of the outcomes are crucial for developing reliable predictive models. It is essential to assess the relevance of the model system for the specific food of interest and the potential limitations during data acquisition and the modeling process. Failure to address these limitations can render the execution of the modeling unfeasible and result in unsatisfactory results [97].

6 Notes

1. *Extrinsic Factors*: These factors are related to the environmental conditions around the analyzed sample, also known as environmental limitation factors. Some examples are the relative humidity, temperature, and gaseous composition of the environment in which this sample is inserted.
2. *Intrinsic Factors*: Intrinsic factors are those related to the sample in which the microorganism is found, thus called substrate limitations. Some examples of this type of factor are composition and availability of nutrients in the medium, pH, redox potential (Eh), barriers and components with antimicrobial action, as well as the water activity of the substrate.
3. *Parameters*: They can be either a value related to a quantity (as the flow and volume parameters are related to the magnitude of a liquid, or the average parameter, which is related to the population quantity) or a variable that can take different values and, as a consequence, change other values of an equation.
4. *Variables*: Variables are generally represented by a letter, which differs from the preestablished values of an equation. They can be divided into dependent (where there are also interdependent) and nondependent (independent). The first group is related to some independent variable, and this second group can also be called a factor. Dependent variables, also known as decision variables, are under the system modeler's control and influence.

References

1. Njage PMK, Sawe CT, Onyango CM et al (2017) Microbial performance of food Safety control and assurance activities in a fresh produce processing sector measured using a microbial assessment scheme and statistical modeling. J Food Prot 80:177–188. https://doi.org/10.4315/0362-028X.JFP-16-233
2. Chaoniruthisai P, Punnakitikashem P, Rajchamaha K (2018) Challenges and difficulties in the implementation of a food safety management system in Thailand: a survey of BRC certified food productions. Food Control 93: 274–282. https://doi.org/10.1016/j.foodcont.2018.06.004
3. Dorota Z, Danuta K-K, Antoni G, Motyl I (2014) Predictive modelling of Lactobacillus casei KN291 survival in fermented soy beverage. J Microbiol 52:169–178. https://doi.org/10.1007/s12275-014-3045-0
4. Ross T, McMeekin TA (1994) Predictive microbiology. Int J Food Microbiol 23:241–264. https://doi.org/10.1016/0168-1605(94)90155-4
5. Van Impe JFM, Poschet F, Geeraerd A, Vereecken KM (2005) Towards a novel class of predictive microbial growth models. Int J Food Microbiol 100:97–105. https://doi.org/10.1016/j.ijfoodmicro.2004.10.007
6. Stavropoulou E, Bezirtzoglou E (2019) Predictive modeling of microbial behavior in food. Foods 8:654. https://doi.org/10.3390/foods8120654
7. Schlundt J, Tay MYF, Chengcheng H, Liwei C (2020) Food security: microbiological and chemical risks. In: Masys AJ, Izurieta R, Reina Ortiz M (eds) Global health security. Advanced sciences and technologies for security applications. Springer, Cham, pp 231–274
8. Bigelow WD, Esty JR (1920) The thermal death point in relation to time of typical thermophilic organisms. J Infect Dis 27(6): 602–617
9. Bigelow WD (1921) The logarithmic nature of thermal death time curves. J Infect Dis 29: 528–536. https://doi.org/10.1093/infdis/29.5.528
10. Esty JR, Meyer K (1922) The heat resistance of the spores of B. botulinus and allied anaerobes. J Infect Dis 31:650–664
11. Ross T, McMeekin TA, Baranyi J (2014) Predictive microbiology and food safety. In: Encyclopedia of food microbiology, 2nd edn. Elsevier, pp 59–68
12. Fávero LP, Belfiore P (2019) Data mining and multilevel modeling. In: Data science for business and decision making, 1st edn. Elsevier, pp 983–1063
13. Valdramidis V (2016) Predictive microbiology. In: Modeling in food microbiology. Elsevier, pp 1–15
14. Perez-Rodriguez F, Valero A (2013) Predictive microbiology in foods. Springer New York, New York, NY
15. Buchanan RL (1993) Developing and distributing user-friendly application software. J Ind Microbiol 12:251–255
16. Dalgaard P, Buch P, Silberg S (2002) Seafood Spoilage Predictor—development and distribution of a product specific application software. Int J Food Microbiol 73:343–349. https://doi.org/10.1016/S0168-1605(01)00670-5
17. Geeraerd AH, Valdramidis VP, Van Impe JF (2005) GInaFiT, a freeware tool to assess non-log-linear microbial survivor curves. Int J Food Microbiol 102:95–105. https://doi.org/10.1016/j.ijfoodmicro.2004.11.038
18. Koseki S (2009) Microbial Responses Viewer (MRV): a new ComBase-derived database of microbial responses to food environments. Int J Food Microbiol 134:75–82. https://doi.org/10.1016/j.ijfoodmicro.2008.12.019
19. González SC, Possas A, Carrasco E et al (2019) 'MicroHibro': a software tool for predictive microbiology and microbial risk assessment in foods. Int J Food Microbiol 290:226–236. https://doi.org/10.1016/j.ijfoodmicro.2018.10.007
20. Baranyi J, Tamplin ML (2004) ComBase: a common database on microbial responses to food environments. J Food Prot 67:1967–1971. https://doi.org/10.4315/0362-028X-67.9.1967
21. Garre A, Fernández PS, Lindqvist R, Egea JA (2017) Bioinactivation: software for modelling dynamic microbial inactivation. Food Res Int 93:66–74. https://doi.org/10.1016/j.foodres.2017.01.012
22. Pla M-L, Oltra S, Esteban M-D et al (2015) Comparison of primary models to predict microbial growth by the plate count and absorbance methods. Biomed Res Int 2015:1–14. https://doi.org/10.1155/2015/365025
23. King T, Cole M, Farber JM et al (2017) Food safety for food security: relationship between global megatrends and developments in food safety. Trends Food Sci Technol 68:160–175. https://doi.org/10.1016/j.tifs.2017.08.014
24. Membré J-M, Boué G (2018) Quantitative microbiological risk assessment in food

industry: theory and practical application. Food Res Int 106:1132–1139. https://doi.org/10.1016/j.foodres.2017.11.025

25. Elliott PH (1996) Predictive microbiology and HACCP. J Food Prot 59:48–53. https://doi.org/10.4315/0362-028X-59.13.48
26. Alvarenga VO, Brito LM, Lacerda ICA (2022) Application of mathematical models to validate emerging processing technologies in food. Curr Opin Food Sci 48:100928. https://doi.org/10.1016/j.cofs.2022.100928
27. Petran RL, Grieme LE, Foong-Cunningham S (2015) 6. Culture methods for enumeration of microorganisms. In: Tortorello YS, Lou M (eds) Compendium of methods for the microbiological examination of foods. American Public Health Association
28. Brehm-Stecher B, Lou TM (2015) 4. Microscopic methods. In: Salfinger Y, Lou TM (eds) Compendium of methods for the microbiological examination of foods. American Public Health Association
29. Powell EO (1956) Growth rate and generation time of bacteria, with special reference to continuous culture. J Gen Microbiol 15:492–511. https://doi.org/10.1099/00221287-15-3-492
30. Horváth S (1970) Importance of the generation time in microbiological experiments. Folia Microbiol (Praha) 15:259–266. https://doi.org/10.1007/BF02869052
31. Peleg M, Corradini MG (2011) Microbial growth curves: what the models tell us and what they cannot. Crit Rev Food Sci Nutr 51: 917–945. https://doi.org/10.1080/10408398.2011.570463
32. Guillier L (2016) Predictive microbiology models and operational readiness. Procedia Food Sci 7:133–136. https://doi.org/10.1016/j.profoo.2016.05.003
33. Liao H, Tang M, Li Z, Lev B (2019) Bibliometric analysis for highly cited papers in operations research and management science from 2008 to 2017 based on Essential Science Indicators. Omega 88:223–236. https://doi.org/10.1016/j.omega.2018.11.005
34. Winston WL (2022) Operations research: applications and algorithms. Cengage Learning
35. Shapiro JF (1993) Mathematical programming models and methods for production planning and scheduling. In: Graves SC, Rinnooy Kan AHG, Zipkin PH (eds) Hanbooks operations research and management science, pp 371–443
36. Smelt JPPM, Brul S (2014) Thermal inactivation of microorganisms. Crit Rev Food Sci Nutr 54:1371–1385. https://doi.org/10.1080/10408398.2011.637645
37. Caldwell J, Ram YM (1999) Mathematical modelling. Springer, Netherlands, Dordrecht
38. Dong Q, Tu K, Guo L et al (2007) Response surface model for prediction of growth parameters from spores of Clostridium sporogenes under different experimental conditions. Food Microbiol 24:624–632. https://doi.org/10.1016/j.fm.2006.12.003
39. Chang Y-J, Wah BW (1994) Polynomial programming using Groebner Bases. In: Proceedings eighteenth annual international computer software and applications conference (COMPSAC 94). IEEE Computer Society Press, pp 236–241
40. Lebert I, Robles-Olvera V, Lebert A (2000) Application of polynomial models to predict growth of mixed cultures of Pseudomonas spp. and Listeria in meat. Int J Food Microbiol 61:27–39. https://doi.org/10.1016/S0168-1605(00)00359-7
41. Albert I, Mafart P (2005) A modified Weibull model for bacterial inactivation. Int J Food Microbiol 100:197–211. https://doi.org/10.1016/j.ijfoodmicro.2004.10.016
42. Lobacz A, Kowalik J (2015) A predictive model for listeria monocytogenes in UHT dairy products with various fat content during cold storage. J Food Saf 35:119–127. https://doi.org/10.1111/jfs.12163
43. Koseki S, Koyama K, Abe H (2021) Recent advances in predictive microbiology: theory and application of conversion from population dynamics to individual cell heterogeneity during inactivation process. Curr Opin Food Sci 39:60–67. https://doi.org/10.1016/j.cofs.2020.12.019
44. Almeida JS (2002) Predictive non-linear modeling of complex data by artificial neural networks. Curr Opin Biotechnol 13:72–76. https://doi.org/10.1016/S0958-1669(02)00288-4
45. Najjar YM, Basheer IA, Hajmeer MN (1997) Computational neural networks for predictive microbiology: I. methodology. Int J Food Microbiol 34:27–49. https://doi.org/10.1016/S0168-1605(96)01168-3
46. Membré J-M, Valdramidis VP (2016) Modeling food microbiology, 1st edn. Elsevier, London
47. McMeekin TA, Olley J, Ratkowsky DA, Ross T (2002) Predictive microbiology: towards the interface and beyond. Int J Food Microbiol 73:395–407. https://doi.org/10.1016/S0168-1605(01)00663-8
48. Baranyi J, Buss da Silva N, Ellouze M (2017) Rethinking tertiary models: relationships between growth parameters of Bacillus cereus

strains. Front Microbiol 8. https://doi.org/10.3389/fmicb.2017.01890

49. Buchanan RL (1993) Predictive food microbiology. Trends Food Sci Technol 4:6–11. https://doi.org/10.1016/S0924-2244(05)80004-4
50. Buchanan R, Whiting R, Damert W (1997) When is simple good enough: a comparison of the Gompertz, Baranyi, and three-phase linear models for fitting bacterial growth curves. Food Microbiol 14:313–326. https://doi.org/10.1006/fmic.1997.0125
51. Zwietering MH, Koos JTDE, Hasenack BE et al (1991) Modeling of bacterial growth function of temperature. Appl Environ Microbiol 57:1094–1101
52. Baranyi J, Roberts TA (1994) A dynamic approach to predicting bacterial growth in food. Int J Food Microbiol 23:277–294. https://doi.org/10.1016/0168-1605(94)90157-0
53. Geeraerd AH, Herremans CH, Van Impe JF (2000) Structural model requirements to describe microbial inactivation during a mild heat treatment. Int J Food Microbiol 59:185–209. https://doi.org/10.1016/S0168-1605(00)00362-7
54. Geeraerd AH, Valdramidis VP, Devlieghere F et al (2004) Development of a novel approach for secondary modelling in predictive microbiology: incorporation of microbiological knowledge in black box polynomial modelling. Int J Food Microbiol 91:229–244. https://doi.org/10.1016/S0168-1605(03)00388-X
55. Ross T (1993) Belehradek-type models. J Ind Microbiol 12:180–189. https://doi.org/10.1007/BF01584188
56. Van Derlinden E, Mertens L, Van Impe JF (2013) The impact of experiment design on the parameter estimation of cardinal parameter models in predictive microbiology. Food Control 29:300–308. https://doi.org/10.1016/j.foodcont.2012.06.018
57. Zwietering MH, Wijtzes T, De Wit JC, Riet KV (1992) A decision support system for prediction of the microbial spoilage in foods. J Food Prot 55:973–979. https://doi.org/10.4315/0362-028X-55.12.973
58. Psomas AN, Nychas G-J, Haroutounian SA, Skandamis PN (2011) Development and validation of a tertiary simulation model for predicting the growth of the food microorganisms under dynamic and static temperature conditions. Comput Electron Agric 76:119–129. https://doi.org/10.1016/j.compag.2011.01.013
59. Psomas AN, Nychas GJ, Haroutounian SA, Skandamis PN (2011) Development and validation of a tertiary simulation model for predicting the growth of the food microorganisms under dynamic and static temperature conditions. Comput Electron Agric 76:119–129. https://doi.org/10.1016/j.compag.2011.01.013
60. Garre A, Koomen J, den Besten H, Zwietering M (2022) biogrowth: Modelling of population growth
61. Garre A, Fernandez JLP-SPS, Egea JA (2019) bioOED: optimum experiment design for microbial inactivation. Dynamic inactivation
62. Garre Perez A, Yeak KY, Pampoukis G et al (2002) D database of microbial inactivation. An innovative online resource for data analysis and meta-regression modeling in predictive microbiology. In: FoodMicro 2022 abstract book: next generation challenges in food microbiology, p 66
63. Food and Drug Administration Center for Food Safety and Applied Nutrition (FDA/CFSAN), Joint Institute for Food Safety and Applied Nutrition (JIFSAN) and Risk Sciences International (RSI) (2021). FDA-iRISK® version 4.2. FDA CFSAN. College Park, Maryland. Available at https://irisk.foodrisk.org/
64. de Alba AM, Buschhardt T, Swaid A et al (2018) FSK-Lab – an open source food safety model integration tool. Microb Risk Anal 10:13–19. https://doi.org/10.1016/j.mran.2018.09.001
65. Dalgaard P (2014) Food spoilage and safety predictor (FSSP) software
66. Petzoldt T (2022) Growthrates: estimate growth rates from experimental data
67. Huang L (2014) IPMP 2013 — a comprehensive data analysis tool for predictive microbiology. Int J Food Microbiol 171:100–107. https://doi.org/10.1016/j.ijfoodmicro.2013.11.019
68. Gonzales-Barron U, Cadavez VAP Listeria Cardinal Models
69. Siqueira AA, de Carvalho PGS, Mendes MLM, Shiosaki RK (2014) MicroFit: um software gratuito para desenvolvimento e ajuste de modelos matemáticos de crescimento bacteriano. Braz J Food Technol 17:329–339. https://doi.org/10.1590/1981-6723.6414
70. Liu Y, Wang X, Liu B et al (2021) Microrisk lab: an online freeware for predictive microbiology. Foodborne Pathog Dis 18:607–615. https://doi.org/10.1089/fpd.2020.2919

71. Baty F, Marie-Laure Delignette-Muller Aurelie Siberchicot (2021) nlsMicrobio: nonlinear regression in predictive microbiology
72. Cadavez V, Kooh P, Sanaa M, Gonzales-Barron U Pathogen in foods database
73. Polese P, Del Torre M, Stecchini ML (2018) Praedicere Possumus: an Italian web-based application for predictive microbiology to ensure food safety. Ital J Food Saf 7. https://doi.org/10.4081/ijfs.2018.6943
74. Leporq B, Membré J-M, Dervin C et al (2005) The "Sym'Previus" software, a tool to support decisions to the foodstuff safety. Int J Food Microbiol 100:231–237. https://doi.org/10.1016/j.ijfoodmicro.2004.10.006
75. McKellar RC, Lu X (2003) Modeling microbial responses in food. CRC Press
76. Mertens L, Van Derlinden E, Van Impe JF (2012) Comparing experimental design schemes in predictive food microbiology: optimal parameter estimation of secondary models. J Food Eng 112:119–133. https://doi.org/10.1016/j.jfoodeng.2012.03.018
77. van Boekel MAJS, Zwietering MH (2007) Experimental design, data processing and model fitting in predictive microbiology. In: Modelling microorganisms in food. Elsevier, pp 22–43
78. Burman E, Bengtsson-Palme J (2021) Microbial community interactions are sensitive to small changes in temperature. Front Microbiol 12. https://doi.org/10.3389/fmicb.2021.672910
79. Collins LM, Dziak JJ, Li R (2009) Design of experiments with multiple independent variables: a resource management perspective on complete and reduced factorial designs. Psychol Methods 14:202–224. https://doi.org/10.1037/a0015826
80. Gunst RF, Mason RL (2009) Fractional factorial design. Wiley Interdiscip Rev Comput Stat 1:234–244. https://doi.org/10.1002/wics.27
81. Ghosh S, Flores A (2013) Common variance fractional factorial designs and their optimality to identify a class of models. J Stat Plan Inference 143:1807–1815. https://doi.org/10.1016/j.jspi.2013.06.008
82. Copeland KAF, Nelson PR (2000) Latin squares and two-level fractional factorial designs. J Qual Technol 32:432–439. https://doi.org/10.1080/00224065.2000.11980029
83. Liau P-H, Huang P-H, Ho J-J, Chen Y-H (2014) Hyper-Graeco-Latin squares and fractional factorial designs. Commun Stat Theory Methods 43:2286–2296. https://doi.org/10.1080/03610926.2013.796986
84. Bevilacqua A, Sinigaglia M (2010) Design of experiments: a powerful tool in food microbiology. Current Research, Technology and Education Topics in Applied Microbiology and Microbial Technology, In
85. Ferreira SL, dos Santos WN, Quintella CM et al (2004) Doehlert matrix: a chemometric tool for analytical chemistry—review. Talanta 63:1061–1067. https://doi.org/10.1016/j.talanta.2004.01.015
86. Sautour M, Rouget A, Dantigny P et al (2001) Application of Doehlert design to determine the combined effects of temperature, water activity and pH on conidial germination of Penicillium chrysogenum. J Appl Microbiol 91:900–906. https://doi.org/10.1046/j.1365-2672.2001.01449.x
87. Antwi M, Theys TE, Bernaerts K et al (2008) Validation of a model for growth of Lactococcus lactis and Listeria innocua in a structured gel system: effect of monopotassium phosphate. Int J Food Microbiol 125:320–329. https://doi.org/10.1016/j.ijfoodmicro.2008.04.014
88. Lianou A, Koutsoumanis KP (2013) Evaluation of the strain variability of Salmonella enterica acid and heat resistance. Food Microbiol 34:259–267. https://doi.org/10.1016/j.fm.2012.10.009
89. Hu M, Gurtler JB (2017) Selection of surrogate bacteria for use in food safety challenge studies: a review. J Food Prot 80:1506–1536. https://doi.org/10.4315/0362-028X.JFP-16-536
90. Guillard V, Couvert O, Stahl V et al (2016) Validation of a predictive model coupling gas transfer and microbial growth in fresh food packed under modified atmosphere. Food Microbiol 58:43–55. https://doi.org/10.1016/j.fm.2016.03.011
91. Urban DL, Minor ES, Treml EA, Schick RS (2009) Graph models of habitat mosaics. Ecol Lett 12:260–273. https://doi.org/10.1111/j.1461-0248.2008.01271.x
92. Hamad H, Al-Smadi A, Ijjeh A (2008) Graphical model validation methods for analog and mixedsignal electronic circuits design. In: 2008 international conference on microelectronics. IEEE, pp 353–356
93. Łobacz A, Żulewska J, Kowalik J (2022) Predictive microbiology and risk analysis. In: da Cruz AG, Ranadheera CS, Nazzaro F, Mortazavian AM (eds) Dairy foods. Elsevier, pp 47–68

94. Ross T (1996) Indices for performance evaluation of predictive models in food microbiology. J Appl Bacteriol 81:501–508. https://doi.org/10.1111/j.1365-2672.1996.tb03539.x

95. Moriasi DN, Arnold JG, Van Liew MW et al (2007) Model evaluation guidelines for systematic quantification of accuracy in watershed simulations. Trans ASABE 50:885–900. https://doi.org/10.13031/2013.23153

96. Draper NR, Smith H (1998) Applied regression analysis. Wiley

97. Allende A, Bover-Cid S, Fernández PS (2022) Challenges and opportunities related to the use of innovative modelling approaches and tools for microbiological food safety management. Curr Opin Food Sci 45:100839. https://doi.org/10.1016/j.cofs.2022.100839

Chapter 2

Methods of Inoculation and Quantification for Collecting Data on Microbial Responses in Foods

Federico Tomasello, Antonio Valero, Andrea Serraino, and Arícia Possas

Abstract

The collection of high-quality data, which are generally obtained through the performance of challenge tests, is essential for building predictive models. In these tests, food samples or modified culture media are artificially contaminated with target microbial cultures and exposed to a series of environmental conditions, mimicking real-life scenarios. A series of aspects including strain selection, inoculum preparation, inoculation method, and enumeration techniques may be considered when performing challenge tests to obtain data for model development. In this chapter, a description of the most applied protocols for data generation is provided. The standardization of methodologies for microbial inoculation and quantification for data generation would benefit the predictive microbiology community, by increasing transparency and reproducibility.

Key words Inoculum level, Microbial quantification, Data gathering, Surrogates

1 Introduction

The development of predictive models is based on observations of microbial responses in foods or culture media systems [1]. When evaluating the effect of intrinsic factors (e.g., pH, a_w, antimicrobial compounds) on the microbial responses in a food product or culture media, different formulations are obtained by adjusting these factors to values set at the experimental designs, by adding salt, organic acids, preservatives, etc. These foods or modified media are then artificially contaminated with the target microorganism and exposed to a series of environmental conditions of interest (e.g., temperature, modified atmosphere) for a given time. These environmental conditions are usually representative of the food production processes, including killing steps such as heat and high-pressure treatments [2], and/or logistic distribution chain steps like storage at retail and household environments.

Verônica Ortiz Alvarenga (ed.), *Basic Protocols in Predictive Food Microbiology*, Methods and Protocols in Food Science,
https://doi.org/10.1007/978-1-0716-3413-4_2,

Collection of high-quality data, which rely mainly on experimental designs and microbiological methods, is essential to build efficient and accurate predictive models [1]. Experimental designs will depend on factors such as the purpose of the model to be developed and the economic resources available to generate data (Chapter 1). Microbiological aspects, which englobe strain selection, inoculum preparation, inoculation, and quantification methods, will depend on a myriad of factors which include the purpose of the study, the limit of detection/quantification of the microbiological analysis technique, and the target microorganism and food product evaluated.

The number of strains, source, and the physiological state of the microorganism are key factors to be considered when preparing inocula for challenge tests to obtain data for model development. Once the strains are chosen, different methods can be applied for sample inoculation, depending on the food type, structure, and physical state. For instance, for food powders such as powder milk, dry inoculation methods have been tested [3], while for solid and liquid foods such as liquid milk, wet inoculation methods have been employed [4]. Moreover, depending on the objective of the study, inoculation of solid foods can be performed on food surfaces by spreading or spraying the inoculums, whereas in other cases, the inoculum can be introduced into the internal part of the product with the aid of syringes or pipettes. Regarding liquid and paste foods, inoculation is usually conducted by homogenizing the product with the inoculums.

The classical plate count techniques have been the most widely applied methods for data collection in predictive microbiology, and to date there are many ISO methods specific for detection and quantification of different microbial groups in foods [1]. Overall, the use of standardized methods for data generation to develop predictive models increases reproducibility and transparency within the predictive microbiology community. Hence, protocols which foster data generation using standardized methods are valuable resources for the modeling process.

This chapter presents a description of the protocols applied for data generation for predictive microbiology model development, including procedures for obtaining both growth and inactivation/survival data.

2 Inoculation Methods

In this section, relevant aspects related to the inoculum used for obtaining microbial data for model development are described, including strain selection, maintenance, inoculum preparation, and levels inoculated.

2.1 Strain Selection

The strains used in challenge tests to generate data for model development should be clearly identified and characterized through previous biochemical, and/or serological, and/or genetic methods [5]. The strains should be selected from the ones isolated from the evaluated food matrices (raw materials, ingredients, and final product), production environment, or clinical, food, and environmental samples involved in outbreaks. In all cases, the following requirements must be met:

- Original sources of the strains must be known.
- Strains must be properly identified and characterized.
- Strains must be available in national or international culture collections for future use.

To allow applying the gamma concept for growth modeling [6], strains for which the cardinal parameters are known (e.g., minimal, optimum, and maximum temperature, pH, and a_w for growth) should be preferred whenever possible [4, 7]. Besides, previous knowledge on the cardinal parameters allows selecting strains which can grow within the ranges of environmental conditions considered in the study.

To account for the variability between strains which can be found in processing environments, a mixture of three to five strains of the same microbial species is usually considered for growth potential studies. However, in growth kinetics studies, to estimate growth parameters in foods (e.g., growth rates), only one strain shall be used per challenge test [5].

Surrogate microorganisms may be used in place of specific pathogens, for example, in case the tests are conducted into a processing facility where for safety reasons it is not possible to use pathogenic microorganisms. An ideal surrogate is a strain that retains all the characteristics of the pathogen except for the virulence and usually is closely related but not necessarily the same species of the target pathogen. If a surrogate microorganism is to be used, preliminary work should be done to characterize the strain before use. Some of the characteristics requested to a surrogate are as follows [8]:

- Nonpathogenic
- Similar behavior to the target pathogen when exposed to parameters such as pH, a_w, and temperature
- Stable and consistent growth characteristics
- Easily prepared to yield high density populations and easily enumerated
- Easily differentiated from background microbiota
- Inactivation characteristics and kinetics that can be used to predict those of the target microorganism
- Genetically stable so that results can be replicated independently of laboratory

Table 1
Examples of modeling studies in which surrogate microorganisms were used

Pathogenic strain	Surrogate	Food/media	Process	Reference
Escherichia coli O157:H7	*Escherichia coli* ATCC 25922	Apple cider	UV treatment	Quintero-Ramos et al. [16]
Escherichia coli O157:H7	*Escherichia coli* ATCC 25922	Apple cider	UV treatment	Duffy et al. [17]
Salmonella enterica	*Enterobacter aerogenes* B199A	Alfalfa seeds	Growth	Liu et al. [18]
Salmonella	*Enterobacter faecium* NRRL B-2354	Peanuts	Heat treatment	Casulli et al. [19]
Salmonella Typhimurium	*Salmonella* Typhimurium LT2 ATCC 700720	Cantaloupe	Growth	Chimbombi et al. [20]
Listeria monocytogenes	*Listeria innocua*	Poultry	Heat treatment	Li et al. [21]
Listeria monocytogenes	*Listeria innocua*	Parsley	Heat treatment	Miller et al., [22]
Listeria monocytogenes	*Listeria innocua*	Fermented sausages	High pressure	Bonilauri et al. [23]
Clostridium botulinum spores	*Clostridium sporogenes* PA 3679 spores	Ground beef	High pressure	Zhu et al. [24]
Clostridium botulinum spores	*Clostridium sporogenes* spores	Nutrient broth	Growth	Khanipour et al. [25]
Clostridium botulinum	*Clostridium sporogenes*	Ground beef	Heat treatment	Hong et al. [26]

Examples of surrogate microorganisms used in predictive microbiology studies are listed in Table 1.

2.2 Strains Maintenance

National and international reference laboratory (e.g., European Union Reference Laboratories) or culture collections (e.g., Spanish Type Culture Collection, American Type Culture Collection; see Table 2) if necessary may provide appropriate strains to the laboratory that runs the test. These strains are usually obtained in the lyophilized form and may be activated following manufacturer's instructions.

The selected strains should be held in the laboratory to enable future testing if required, keeping them in culture broth supplemented with glycerol (20–25%) or using cryopreservation beads immersed in cryo-solutions to prevent damage while freezing and to improve the survival during long term storage of microorganisms. In all cases, the selected strains must be stored at temperatures ≤−20 °C.

Table 2
Examples of culture collections worldwide

Culture collection	Website
American Type Culture Collection (ATCC)	https://www.atcc.org
Spanish Type Culture Collection (CECT)	https://www.uv.es/uvweb/spanish-type-culture-collection/en/spanish-type-culture-collection-1285872233521.html
Culture Collection University of Goteborg (CCUG)	https://www.ccug.se
Collection de L'Institut Pasteur (CIP)	https://www.pasteur.fr/en/public-health/biobanks-and-collections/collection-institut-pasteur-cip
National Collections of Industrial Food and Marine Bacteria (NCIMB)	https://www.ncimb.com
China General Microbiological Culture Collection Center (CGMCC)	http://www.cgmcc.net

2.3 Inoculum Preparation

Standardization of the preparation of the inoculum is particularly important to be able to inoculate the food matrix at the expected/desired concentration.

For the preparation of the vegetative cell suspensions, two successive cultures of the chosen strains shall be conducted. The first culture may be conducted in a medium under conditions to enable optimal growth of the strains, reaching the end of the exponential phase or the early stationary phase. The second culture is meant to adapt the strains to the natural conditions of the food, so it should be conducted in a medium mimicking those conditions (at least temperature, pH, a_w) again until the end of the exponential phase or the early stationary phase.

In case of fate studies (growth potential studies), if inoculation is performed in end products to evaluate microbial behavior during their storage at different conditions, it is convenient to consider the adaptation of bacterial cells to stress conditions that may take place during the production processes of the selected products (e.g., HPP treatment as part of inoculum preparation). To this end, strains shall be subjected to treatments that mimic food production processes, e.g., injury and/or stress, to induce adaptation. Examples are treatments with hydrochloric acid to damage bacterial cells and heat, cold, or lactic acid stress [5]. If growth kinetics are being studied, there is no need to induce any adaptation on the selected strain [5].

After this adaptation step, centrifugation and resuspension in a nongrowth promoting diluent (e.g., phosphate buffer solution) is needed to avoid adding nutrients that could interfere with the microbial behavior [5].

In case of using a multi-strain cocktail, aliquots of grown cultures of each individual strain selected, stress-adapted or not, should be transferred to a sterile tube, and each strain must be present at the same concentration on the final cocktail.

The subcultures of the strains selected for the test should be carried out in advance, and the total volume of the inoculum suspension should be large enough to inoculate all the units with a inoculum – sample ratio not exceeding 1:100 [5].

2.4 Inoculum Level

The inoculum concentration used depends mainly on the objective of the study, namely, to determine growth or inactivation/survival of the microorganisms. Moreover, the inoculum concentration shall be justified and adjusted in accordance with the expected microbial behavior in the food matrix. Usually, for challenge tests in which microbial growth in the food matrix is expected, an inoculum level ranging from 10^2 to 10^3 colony-forming units (CFU)/g or mL is used. It is important to consider that the product formulation could lead to a death of the strains before their adaptation and that also an excessive concentration could overwhelm the stability of the product leading to an incorrect conclusion about the growth potential. Thus, these aspects should be considered when selecting the inoculum concentration.

Conversely, in fate studies where microbial survival or inactivation patterns are expected (e.g., heat treatments), the inoculum level shall be increased up to 10^5–10^7 CFU/g or mL to enable the proper evaluation and characterization of the microbial reduction trend during the test, avoiding reaching, in samples, microbial concentrations below the detection limit over the period considered. In any case, the inoculum concentration should be within the quantification limit of the enumeration method used. Determination of microbial suspension counts shall be carried out using, for example, the plate count method, the optical density at 420–660 nm, or the McFarland turbidity standards [9], to aid in calculating the dilutions necessary to achieve the target inoculum in the product (Fig. 1) [5, 8].

2.5 Inoculation Techniques

The aim of this step is to achieve the required level of microorganism in each test unit without modifying the physicochemical properties of the matrix (e.g., pH, a_w). Before inoculation, food samples shall be adapted to the initial temperature of the test. The matrix can be inoculated on the surface, mimicking contamination during production process (e.g., slicing of smoked salmon), or in-depth for homogeneous foods like ground food or creams or for food prepared mixing different ingredients (e.g., mixed salad).

There are various inoculation methods that can be applied depending on the type of product being evaluated. For example, in liquid matrices with high a_w (>0.96), the inoculum could be directly inoculated with mixing, using a negligible amount of

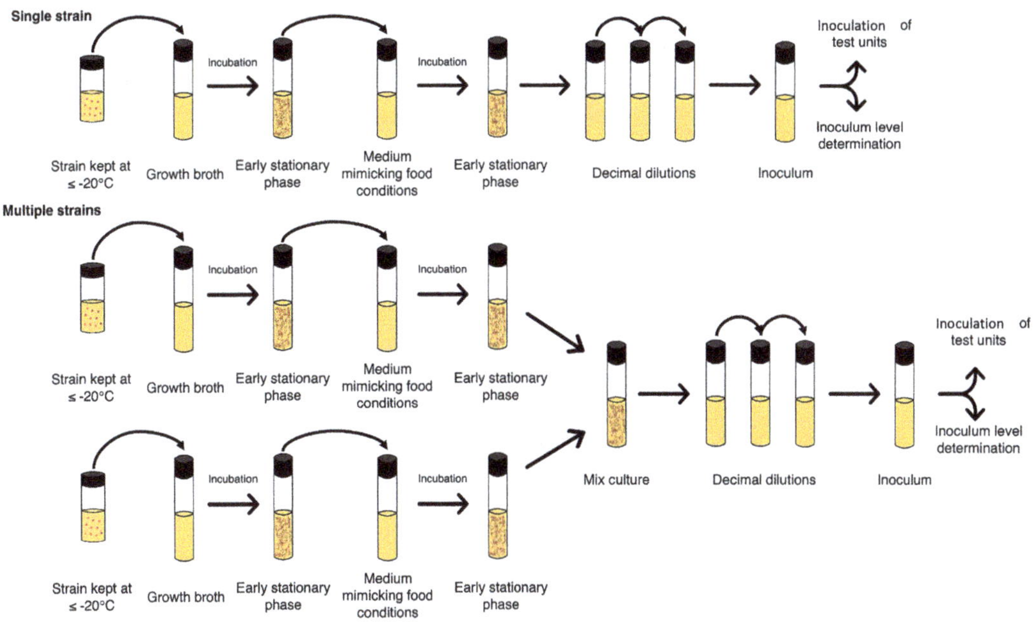

Fig. 1 Strain activation and inoculation preparation

carrier, and adjusting its a_w to that of the matrix. In solid matrices with high a_w (>0.96), such as fresh meat or fish, the use of an atomizer represents a good option. In this technique, the inoculum is suspended in water or buffer and sprayed into the ground product or on the surface. This procedure should be done using devices to ensure personal safety (e.g., cabinet) and using the minimum amount of liquid possible. It is also possible to inoculate the sample using sterile pipettes, distributing the inoculum on the surface or into the ground product (e.g., in cases where it is not possible to ensure the personal safety when spraying the inoculum) [8].

For products with lower a_w (<0.92), there are two options for inoculation. The first uses an inoculum in aqueous form; an example is the atomizer technique, above described paying special attention to using a minimal volume of the carrier solution, also applying a drying period after inoculation, or mixing the wet inoculum with a small amount of food sample, called a "seed" inoculum, which is then mixed with a larger food sample. The second option is the dry inoculation, suggested for products with even lower a_w (≤0.65), in which the inoculum can be prepared by suspension of the challenge microorganisms in water or buffer and adding it to sterile flour or powdered product and allowing it to dry until it reaches extremely low a_w (≤0.5). Alternatively, drying techniques such as the use of vacuum desiccator or freeze dryer, spray dryer, or fluidized bed dryer can be applied to the microbial culture, adding protective agents (e.g., sucrose, skim milk) to increase the microbial survival. The dried inoculum then can be added to the product, under agitation for an even distribution. Despite the longer and more labor intensive procedure, dry inoculums allow a more uniform distribution of microorganism, with minimal impact on the properties of the food matrices [8, 10].

Moreover, product can be inoculated through the packaging using a needle and some type of self-sealing rubber or silicon septum [11]. The package atmosphere is an important extrinsic factor to consider when inoculating food products. Ideally, the product should be first inoculated and then packaged under a condition that best mimics the ones used during commercial production (e.g., vacuum packaging, modified atmosphere packaging (MAP)), including the use of the exact gas mix used for MAP, same packaging material, and similar vacuum levels. However, in some cases, the inoculation previously to packaging is not possible, and the use of syringe for inoculation has been reported in different studies.

3 Controls

Controls are essentials to assist in the proper microbial data interpretation and to understand the influence of the evaluated factors on microbial evolution. They can be used to evaluate relevant physicochemical properties like pH, a_w, organic acid concentration, concentrations of other preservatives, background microbiota, and/or added starter cultures.

Two kinds of controls are usually used: control units and food controls. For the preparation of control units, the same inoculation procedure applied for the test units may be performed using a nongrowth promoting diluent, free of the pathogenic cells. It is recommended to analyze control units at least at the first (t_0) and the last (t_{end}) sampling point of the fate study, although it is better to analyze at least one control unit per sampling point. Food controls on the other hand are non-inoculated food units and are used to verify if the inoculation process modifies the physicochemical properties of the food and/or to evaluate whether the food samples are naturally contaminated with the evaluated microorganism or the background microbiota of the food matrices.

4 Replicates and Number of Units

Depending on the variability of the production process and food characteristics, i.e., intrinsic (e.g., pH, a_w, preservatives), extrinsic (e.g., gas composition), and microbiological properties, a different number of replicates, using different batches, shall be used in the challenge study. The batch variability should be representative of the production process, based on historical data.

If the variability between different batches in the process is sufficient ($\Delta\varphi_{\mathrm{pH}a\mathrm{w}}$ or $\Delta\psi > 0.2$) to produce differences in microbial behavior, more than one batch is necessary to account for variability in microbial responses [12]. In this case, at least three different batches should be used.

For growth rate studies, the "Inter-Batch Physico-Chemical Variability calculator" (URL: http://standards.iso.org/iso/20976/-1/ed-1/en) provided with ISO 20976-1:2019 can be used to assess the impact of inter-batch variability. Moreover, a detailed description of inter-batch variability assessment, based on pH and a_w, is described within the ISO 20976-1:2019. The use of one batch shall be justified (e.g., using the calculator, using a "worst case" batch). Replicates should be independent trials using different batches of product and inoculum to account for variations in product, inoculum, and other factors [4, 5, 11].

The number of analytical units (AU) greatly depends on the experimental design, mainly the duration, since a representative sampling regime must be covered. The sampling interval, and so the number of AU, should be defined based on prior experience with related products and considering the likely duration of survival or rate of growth or inactivation and the product shelf-life. Depending on the product characteristics and expected outcomes for products with a long shelf-life, it may be appropriate to test on a more frequent basis early in the study (e.g., daily) and at longer intervals later in the study. Anyway, a minimum of five to seven points are required to obtain an accurate description of the studied microbial behavior.

A minimum of two AU shall be analyzed at each time interval, during processing or storage, and if possible, analysis of three or more AU shall be preferred.

Sources of intra-batch variability need to be considered when selecting the number of AU. Generally, the number of samples and replicates should be increased in situations of higher variability or uncertainty, but in cases where data from other studies exist, the need for replication may be reduced. Moreover, statistical experimental design (e.g., power analysis) can improve the validity of the study [4, 5, 11].

5 Microbial Enumeration

Microbial quantification may be achieved in many ways: using solid or liquid media, by real-time polymerase chain reaction, etc. In this chapter, we'll cover just the microbial enumeration by plating in solid media, being this the one suggested for general quantification test by the ISO 7218:2007/Amd.1:2013 [13].

Table 3
Examples of enumeration methods by the International Organization for Standardization (ISO)

Microorganism	Enumeration method
Enterobacteriaceae	ISO 21528-2:2017 [27]
Aerobic mesophilic bacteria	ISO 4833-1:2013 [28] ISO 4833-2:2013/Corr.1:2014 [29]
Coagulase-positive staphylococci	ISO 6888-1:2021 [30]
Clostridium perfringens	ISO 7937:2004 [31]
Lactic acid bacteria	ISO 15214:1998 [32]
Thermotolerant *Campylobacter*	ISO 10272-2:2017 [33]
Listeria spp.	ISO 11290-2:2017 [34]
Salmonella spp.	ISO 6579-1:2017 [35] ISO 6579-1:2017/AMD 1:2020 [36]
Escherichia coli	ISO 16649-1:2018 [37]
Yeast and moulds	ISO 21527-1:2008 [38] ISO 21527-2:2008 [39]

For microbial detection and quantification, internationally accepted and validated protocols for the specific microorganism shall be applied (see examples in Table 3).

5.1 Sample Preparation

Quantification of microorganisms is usually performed using AU of 25 g (25 mL) of the sample, and a minimum of 10 g (10 mL) is suggested according to ISO 6887-1:2017 [14]. Before withdrawing the AU, the sample should be well homogenized to ensure the representativity of the portion withdrawn.

The subsequent step is to dilute the AU with the appropriate diluent medium before plating into culture media. The recommended initial dilution is 1:10, obtained by adding n grams or milliliters of AU to $9 \times n$ mL of diluent (e.g., 25 g/mL into 225 mL).

In the case of solid foods, a homogenization step is necessary to obtain a suspension that can be used for the subsequent analysis. Usually, the AU is transferred into a sterile bag, and an amount of diluent necessary to obtain a 1:10 dilution is added and homogenized by agitation in a peristaltic homogenizer (stomacher) for 1–2 min (soft or pasty foods, ground or minced foods, poorly soluble powders). Alternatively, for hard foods, homogenization can be done using sterile jars in a blender avoiding excessive heating of the AU. For liquid foods, this homogenization process is not necessary since it is possible to use the food matrix as it is for the

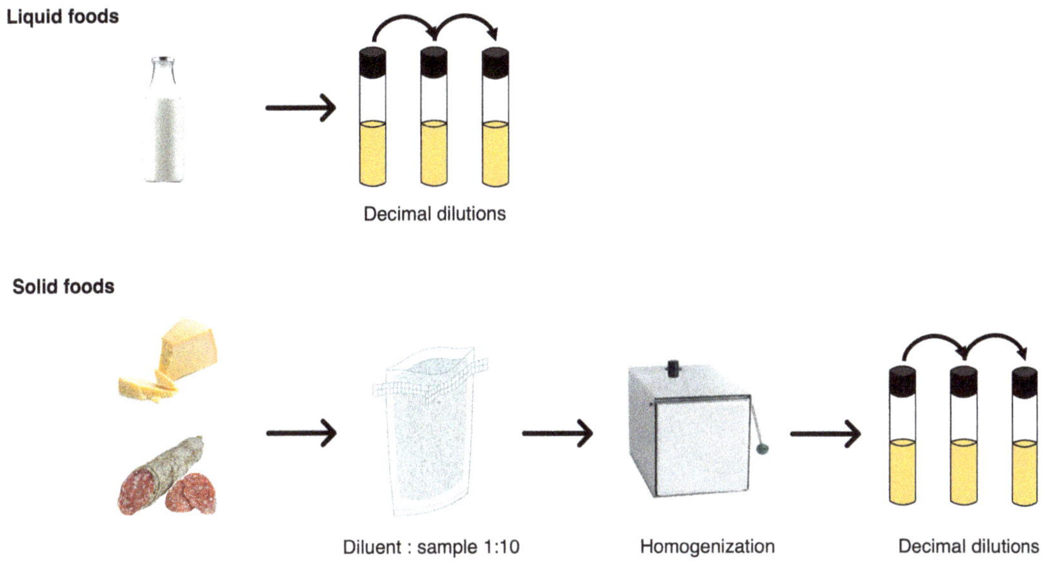

Fig. 2 Scheme of sample preparation for microbiological analysis

analysis; anyway, in case of viscous fluids, it can be achieved by agitation in a flask containing the amount of diluent necessary for a 1:10 dilution.

In cases the number of microorganisms in the AU is expected to be high, a series of tenfold dilutions to reduce the number of cells per unit of volume are necessary to allow quantification. The number of necessary dilutions depends on the expected microbial load and should allow counting between 25–30 and 250–300 CFU per plate [14]. Dilutions are conducted using saline peptone water or buffered peptone water. An example of this procedure is shown in Fig. 2.

5.2 Enumeration Techniques

Depending on the physiological characteristics of the microorganisms, expected levels to be encountered in the AU, and the limit of quantification, different microbial enumeration techniques can be employed. The most applied ones are the pour and the spread plate techniques (Fig. 3).

5.2.1 Pour Plate Technique

This technique has a detection limit of 10 CFU/g for solid and 1 CFU/mL for liquid foods, when 1 mL aliquots of the homogenized AU are used for plating. Inoculating greater volumes distributed over several Petri dishes (up to 2 mL per 90 mm Petri dish) allows to achieve a detection limit of 1 CFU/g for solid products.

The molten culture media should be rapidly cooled down and kept in a water bath at a temperature of 44–46 °C until use. Attention should be paid when removing the tempered agar

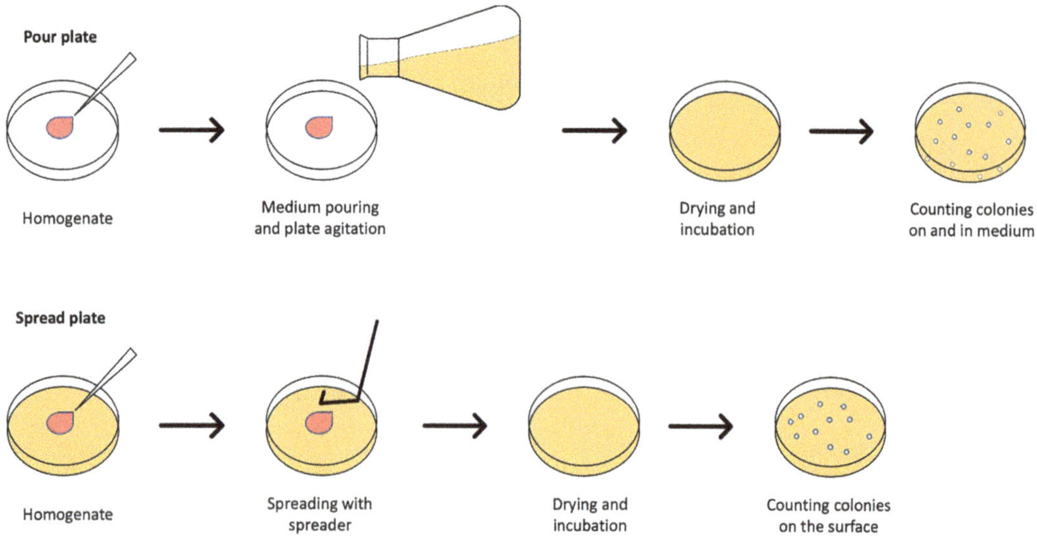

Fig. 3 Techniques for microbial enumeration

medium from the water bath, drying the bottle with a clean towel to prevent water from contaminating the plates. The molten agar medium at 44–46 °C must be poured into each Petri dish containing the AU homogenate (generally, 18–20 mL of agar in 90 mm Petri dishes to obtain at least 3 mm thickness) avoiding excessive agitation of the medium to prevent bubble formation. The molten medium may not be poured directly onto the aliquot of diluted AU. The time elapsed between pipetting the aliquots of diluted AU and pouring the agar media shall not exceed 15 min, to avoid aggregation of colonies. Immediately after pouring, the molten medium and the aliquot shall be carefully mixed to obtain a homogeneous distribution of the microorganisms within the medium, e.g., by gently moving the dish backward and forward, from side to side, and in a circular direction, and allowing it to cool and solidify placed on a cool horizontal surface. Incubation times, temperatures, and atmospheres required by each microorganism can be found in the specific ISO methods (Table 3) [13].

5.2.2 Surface Plate Technique

This standard procedure has a detection limit of 100 CFU/g for solids and 10 CFU/mL for liquids but can be adapted, if necessary, to a detection limit of 10 CFU/g for solid products or 1 CFU/mL for liquid products, by inoculating 1 mL of the initial suspension onto the surface of three different 90 mm Petri dishes. For this method, the use of pre-poured plates with agar medium of at least 3 mm of thickness, level and free from air bubbles and surface moisture, is required.

The standard procedure uses 0.1 mL of the appropriate dilutions of the AU, which should be inoculated into the agar plates using a sterile pipette. Glass or plastic spreader (Drigalski) is usually used to spread the inoculum onto the entire surface of the medium as fast as possible, without touching the sidewalls of the Petri dish and ensuring a uniform distribution over the surface until all aliquot is absorbed by the medium. Finally, plates must be dried before incubation, with the lid on for 15 min at room temperature [13].

5.3 Calculation and Expression of Results

ISO 7218:2007/Amd.1:2013 requires using one plate per dilution, of at least two successive dilutions, or two plates per dilution if only one dilution is used. The use of two plates per dilution may also be applied to improve reliability, and it is mandatory for laboratories that do not operate under quality assurance principles. If two or more dilutions are used, the count at a given dilution should be approximately 10% of the previous/following, with an upper limit of 15.6% and a lower limit of 5.2% [15].

The maximum number of colonies acceptable for enumeration should not exceed 300 in 90 mm Petri dishes, while the minimum number of colonies shall not be less than 25. When dishes with a diameter different from 90 mm are used, the maximum number of colonies shall be increased or decreased in proportion to the surface area of the dishes.

The general rule for counting total or typical colonies reported in ISO 7218:2007/Amd.1:2013 is described by Eq. 1.

$$\text{CFU/g or CFU/mL} = \frac{\Sigma C}{v \times [n_1 + (0.1 \times n_2)] \times d} \tag{1}$$

where

ΣC = sum of colonies (or typical colonies) counted on the plates selected for counting from two successive dilutions

v = volume of the diluted homogenate on each plate (usually 0.1 mL for spread plating or 1 mL for pour plating)

n_1 = number of plates counted from the first dilution selected (usually 1 if no replicate was made or 2 when duplicate was made)

n_2 = number of plates counted from the second dilution selected

d = first dilution retained for counting ($10^0 = 1$, $10^{-1} = 0.1$, $10^{-2} = 0.01$)

The ISO 7218:2007/Amd.1:2013 also reports other ways of counting CFU in unusual situations, such as plates with less than 10 or more than 300 colonies, and establishes exponential notations for the presentation of the results.

Acknowledgments

Project PROBIOFISH (P18 RT 3177): Development and application of predictive microbiology models in minimally processed and ready-to-eat aquaculture products through the application of microorganisms with probiotic and bioprotective potential. The project is funded by the Consejería de Economía, Conocimiento, Empresas y Universidad, Junta de Andalucía.

References

1. Perez-Rodriguez F, Valero A (2013) Predictive microbiology in foods. Springer New York, New York, NY
2. Possas A, Pérez-Rodríguez F, Valero A et al (2018) Mathematical approach for the *listeria monocytogenes* inactivation during high hydrostatic pressure processing of a simulated meat medium. Innov Food Sci Emerg Technol 47: 271–278
3. Wiertzema JR, Borchardt C, Beckstrom AK et al (2019) Evaluation of methods for inoculating dry powder foods with Salmonella enterica, Enterococcus faecium, or Cronobacter sakazakii. J Food Prot 82:1082–1088
4. Bergis H, Bonanno L, Asséré A, et al (2021) EURL Lm Technical guidance document on challenge tests and durability studies for assessing shelf-life of ready-to-eat foods related to *Listeria monocytogenes*. http://eurl-listeria.anses.fr
5. International Organization for Standardization (ISO) (2019) ISO 20976-1:2019. Microbiology of the food chain – requirementes and guidelines for conducting challenge tests of food and feed products – Part 1: challenge tests to study growth potential, lag time and maximum growth rate. https://www.iso.org/standard/69673.html
6. Zwietering MH, Wijtzes T, Rombouts FM et al (1993) A decision support system for prediction of microbial spoilage in foods. J Ind Microbiol 12:324–329
7. International Organization for Standardization (ISO) (2022) ISO 23691. Microbiology of the food chain — determination and use of cardinal values. https://www.iso.org/standard/76665.html?browse=tc
8. U.S. Food and Drug Administration (2003) Microbiological challenge testing. Compr Rev Food Sci Food Saf 2:46–50
9. Koch AL (2014) Growth measurement. In: Methods for general and molecular microbiology. Wiley, pp 172–199
10. Xu J, Song J, Tan J et al (2020) Dry-inoculation methods for low-moisture foods. Trends Food Sci Technol 103:68–77
11. National Advisory Committee on Microbiological Criteria for Foods (2010) Parameters for determining inoculated pack / challenge study protocols. J Food Prot 73:140–202
12. International Organization for Standardization (ISO) (2019) User's guide of the inter-batch physico-chemical variability calculator
13. International Organization for Standardization (ISO) (2013) ISO 7218:2007/Amd.1:2013. Microbiology of food and animal feeding stuffs — general requirements and guidance for microbiological examinations-Amendment 1. https://www.iso.org/standard/52204.html
14. International Organization for Standardization (ISO) (2017) ISO 6887-1:2017 Microbiology of the food chain — preparation of test samples, initial suspension and decimal dilutions for microbiological examination — Part 1: general rules for the preparation of the initial suspension and decimal dilutions. https://www.iso.org/standard/63335.html
15. International Organization for Standardization (ISO) (2005) ISO 14461-2:2005. Milk and milk products — quality control in microbiological laboratories — Part 2: determination of the reliability of colony counts of parallel plates and subsequent dilution steps. https://www.iso.org/standard/36838.html
16. Quintero-Ramos A, Churey JJ, Hartman P et al (2004) Modeling of *Escherichia coli* inactivation by UV irradiation at different pH values in apple cider. J Food Prot 67:1153–1156
17. Duffy S, Churey J, Worobo RW et al (2000) Analysis and modeling of the variability associated with UV inactivation of *Escherichia coli* in apple cider. J Food Prot 63:1587–1590
18. Liu B, Schaffner DW (2007) Mathematical modeling and assessment of microbial migration during the sprouting of alfalfa in trays in a nonuniformly contaminated seed batch using

Enterobacter aerogenes as a surrogate for *Salmonella* Stanley. J Food Prot 70:2602–2605

19. Casulli KE, Igo MJ, Schaffner DW et al (2021) Modeling inactivation kinetics for *Enterococcus faecium* on the surface of peanuts during convective dry roasting. Food Res Int 150:110766
20. Chimbombi E, Moreira RG, Castell-Perez EM et al (2013) Assessing accumulation (growth and internal mobility) of *Salmonella* Typhimurium LT2 in fresh-cut cantaloupe (*Cucumis melo* L.) for optimization of decontamination strategies. Food Control 32:574–581
21. Li M, Pradhan A, Cooney L et al (2011) A predictive model for the inactivation of *Listeria innocua* in cooked poultry products during postpackage pasteurization. J Food Prot 74: 1261–1267
22. Miller FA, Ramos BF, Gil MM et al (2011) Heat inactivation of *Listeria innocua* in broth and food products under non-isothermal conditions. Food Control 22:20–26
23. Bonilauri P, Merialdi G, Ramini M et al (2021) Modeling the behavior of *Listeria innocua* in Italian salami during the production and high-pressure validation of processes for exportation to the U.S. Meat Sci 172:108315
24. Zhu S, Naim F, Marcotte M et al (2008) High-pressure destruction kinetics of *Clostridium sporogenes* spores in ground beef at elevated temperatures. Int J Food Microbiol 126:86–92
25. Khanipour E, Flint SH, McCarthy OJ et al (2016) Modelling the combined effects of salt, sorbic acid and nisin on the probability of growth of *Clostridium sporogenes* in a controlled environment (nutrient broth). Food Control 62:32–43
26. Hong YK, Huang L, Yoon WB (2016) Mathematical modeling and growth kinetics of *Clostridium sporogenes* in cooked beef. Food Control 60:471–477
27. International Organization for Standardization (ISO) (2017) ISO 21528-2:2017. Microbiology of the food chain — horizontal method for the detection and enumeration of Enterobacteriaceae — Part 2: colony-count technique. https://www.iso.org/standard/36588.html
28. International Organization for Standardization (ISO) (2013) ISO 4833-1:2013 Microbiology of the food chain — horizontal method for the enumeration of microorganisms — Part 1: colony count at 30 °C by the pour plate technique. https://www.iso.org/standard/53728.html
29. International Organization for Standardization (ISO) (2014) ISO 4833-2:2013/COR 1: 2014. Microbiology of the food chain — horizontal method for the enumeration of microorganisms — Part 2: colony count at 30 °C by the surface plating technique — Technical Corrigendum 1. https://www.iso.org/standard/65999.html
30. International Organization for Standardization (ISO) (2021) ISO 6888-1:2021. Microbiology of the food chain — horizontal method for the enumeration of coagulase-positive staphylococci (*Staphylococcus aureus* and other species) — Part 1: method using Baird-Parker agar medium. https://www.iso.org/standard/76672.html
31. International Organization for Standardization (ISO) (2004) ISO 7937:2004. Microbiology of food and animal feeding stuffs — horizontal method for the enumeration of *Clostridium perfringens* — colony-count technique. https://www.iso.org/standard/36588.html
32. International Organization for Standardization (ISO) (1998) ISO 15214:1998. Microbiology of food and animal feeding stuffs — horizontal method for the enumeration of mesophilic lactic acid bacteria — colony-count technique at 30 degrees C. https://www.iso.org/standard/26853.html
33. International Organization for Standardization (ISO) (2017), ISO 10272-2:2017. Microbiology of the food chain — horizontal method for detection and enumeration of *Campylobacter* spp. — Part 2: colony-count technique. https://www.iso.org/standard/63228.html
34. International Organization for Standardization (ISO) (2017) ISO 11290-2:2017. Microbiology of the food chain — horizontal method for the detection and enumeration of *Listeria monocytogenes* and of *Listeria* spp. — Part 2: enumeration method. https://www.iso.org/standard/60314.html
35. International Organization for Standardization (ISO) (2017) ISO 6579-1:2017. Microbiology of the food chain — horizontal method for the detection, enumeration and serotyping of *Salmonella* — Part 1: detection of *Salmonella* spp. https://www.iso.org/standard/56712.html
36. International Organization for Standardization (ISO) (2020) ISO 6579-1:2017/AMD 1: 2020. Microbiology of the food chain — horizontal method for the detection, enumeration and serotyping of *Salmonella* — Part 1: detection of *Salmonella* spp. — Amendment 1: broader range of incubation temperatures, amendment to the sta. https://www.iso.org/standard/76671.html

37. International Organization for Standardization (ISO) (2018) ISO 16649-1:2018. Microbiology of the food chain — horizontal method for the enumeration of beta-glucuronidase-positive *Escherichia coli* — Part 1: colony-count technique at 44 degrees C using membranes and 5-bromo-4-chloro-3-indolyl beta-D-glucuronide. https://www.iso.org/standard/64951.html
38. International Organization for Standardization (ISO) (2008). ISO 21527-1:2008. Microbiology of food and animal feeding stuffs — horizontal method for the enumeration of yeasts and moulds — Part 1: colony count technique in products with water activity greater than 0.95. https://www.iso.org/standard/38275.html
39. International Organization for Standardization (ISO) (2008) ISO 21527-2:2008. Microbiology of food and animal feeding stuffs — horizontal method for the enumeration of yeasts and moulds — Part 2: colony count technique in products with water activity less than or equal to 0.95. https://www.iso.org/standard/38276.html

Chapter 3

The Influence of Food Matrices on Microbial Growth

Peter Myintzaw and Michael Callanan

Abstract

The effects of water availability, acidity, and food structure on the growth of food safety and spoilage microbes can be predicted from experiments in laboratory media, but these predictions often need to be validated in food matrices. Generating accurate and reproducible growth data in food matrices presents particular challenges when compared to experiments in laboratory media. The aim of this chapter is to provide the reader with an introduction and guidelines on measuring the influence of intrinsic food properties on microbial growth rates in real food matrices.

Key words Food structure, Culture medium, Food properties, Liquid models

1 Introduction

The growth, survival, and inactivation of microorganisms have largely been researched using a deliberately adjusted liquid culture medium to mimic a true food matrix. This data has been critical in the progression of food safety management from a culture of end product and challenge testing to a risk management based approach. In risk management, detailed knowledge of a critical risk such as the growth of a foodborne microbial pathogen is essential as any implemented control measures are reliant on an accurate and comprehensive risk profile. However, many studies have emphasized the disparities between predictions made in a laboratory broth model and those made in a real food matrix [1–3]. The pH, water activity (a_w), oxidation-reduction potential (*Eh*), nutrient content, and the presence of antimicrobial constituents are the most critical aspects of the food matrix that influence microbial proliferation. These intrinsic factors can be naturally existing or intentionally altered to enhance food safety. Combining one or more of these "hurdles" has been demonstrated to be effective in suppressing microbial growth while retaining the food's organoleptic quality. In addition to intrinsic factors, properties of food matrix

Verônica Ortiz Alvarenga (ed.), *Basic Protocols in Predictive Food Microbiology*, Methods and Protocols in Food Science,
https://doi.org/10.1007/978-1-0716-3413-4_3,

such as the structure of food itself, presence of higher fat content, or state of gelation can also have a major influence on microbial growth [1, 4–7]. The obvious comparison is between a whole carrot stick and carrot puree; both products have the same intrinsic properties, but the puree has less structural components, allowing for better microbial proliferation than the whole carrot.

The food industry and regulators are moving away from end product testing to relying on quantitative microbial risk assessment (QMRA) based approaches to food safety. Predictive microbiology (PM) is an integral aspect of any QMRA approach. PM is a concept that integrates microbiology, mathematics, and statistics and constructs mathematical models to infer microbial behavior within food environmental conditions [8–10]. PM is widely used to evaluate microbial proliferation, growth limitations, and inactivation rate in a matrix in order to develop new products and processes, product reformulation, and determination of storage conditions and shelf life. However, PM tools such as ComBase (www.combase.cc) relies significantly on data generated in laboratory media due to the ease and reproducibility of growth rate data generated under controlled conditions. Laboratory media also facilitate growth measurement using optical density changes which is more rapid and less intensive than other methods but cannot be used in foods where direct counting methods are required.

1.1 Microbial Growth Quantification

Direct microscopic count (DMC) was the first method employed for quantitative enumeration of both spoilage and pathogenic microbes in the food industry [11]. However, the limitation of not being able to distinguish living or dead cells [12] meant viable plate counts (VPC) quickly became the gold standard [13, 14] and has legal recognition for routine enumeration and identification of microbes in food samples [15–17]. VPC and derived methods (Table 1) count living bacteria that can grow as colonies on selective and nonselective solid media after incubation. While a single colony may consist of multiple bacterial cells, the assumption is made that a colony originates from a single cell. Other methods such as flow cytometry can determine cells as active and/or total units and are an accepted alternative to conventional plate count enumeration [17, 19]. The principle involves generating a liquid suspension of a sample containing microbes and adding it into a fluid stream and then scanning the cells by intercepting laser beams as they pass past the observation point [20]. For more information *see* Chapter 2.

Alternative rapid and indirect detection methods for foods have been developed with the intent of finding a speedy, sensitive, specific, and cost-effective process to replace time-consuming and laborious VPC methods. Piezoelectric crystals; impedimentary, redox reactions; and calorimetry have all been used to measure physical parameters, as well as the detection of cellular components such as ATP (by bioluminescence), DNA, protein, and lipid

Table 1
Direct microbial growth detection and quantification methods

Method	Synonyms	Description	Pro	Con	Reference
Direct microscopic counts	Direct count	Aliquot sample in Petroff-Hausser counting chamber for viewing using a microscope	Rapid, simple	Counts both dead and live cell; only higher cell concentration can be counted	(Huhtanen & Jones, [18])
Viable plate count	Aerobic platecount (APC), total plate count (TPC), standard plate count, mesophilic count, heterotrophic plate count (HPC)	Serially diluted sample plated on an appropriate nutrient medium (agar plate) becomes visible to the naked eye	Sensitive, only counts viable bacteria, possible to use selective or nonselective medium	Laborious, specific medium required to detect target microbes, may require up to 72 h or more to get result	Diez-Gonzalez, [14]
3 M™ Petrifilm™ Plate method	N/A	Similar to plate count	Reduces labor, workspace, and waste	Hydrolysable microbe gelling agent in the Petrifilm and higher salt and pH may have effect on result	Diez-Gonzalez, [14]
SimPlate Total plate count method	N/A	The medium/sample mixture is dispensed into a SimPlate device and incubated for 24–28 h; subsequently, total aerobic plate count is determined by counting the fluorescent wells	Shorter incubation time, higher number counting range	Endogenous enzyme activities that react with the indicators in the SimPlate medium can cause false-positive reactions	Diez-Gonzalez, [14]
Hydrophobic grid membrane filter method	HGMF	The sample is filtered through the HGMF and then incubated on a specially prepared agar plate. In HGMF, the colonies develop and absorb a green dye from the agar medium	Improved sensitivity compared to APC; 10,000 cells per sample can be counted	Laborious, the quick green dye may be stained by some food (e.g., corn and tuna)	Diez-Gonzalez, [14]

(continued)

Table 1 (continued)

Method	Synonyms	Description	Pro	Con	Reference
Plate loop count method	N/A	A calibrated inoculating loop with a continuous pipetting syringe is used to distribute sample onto a Petri dish, which is then mixed with molten agar, and colony counts are performed same as agar pour-plate techniques	Quick and easy to process large numbers of samples, reduces workload, and material	Homogenous, non-particulate samples only	Diez-Gonzalez, [14]
TEMPO® TVC automated total vial count	N/A	This utilizes a miniaturized version of the most probable number (MPN) technique combined with the detection of growth by fluorescence of a proprietary chromophore resulting from microbial activity	It is possible to analyze as many as 500 samples per day. Reduces labor cost, ease of use	Requires numbers of instrumentation	Diez-Gonzalez, [14]

derivatives (by biochemical methods) [21]. However, VPC methods remain the gold standard for determining cell numbers in food matrices for accurate and reliable growth data generation.

1.2 Food Structure

Typically, the growth potential of the microbes in a food matrix has been tested by inoculating organism of concern in a particular food matrix which likely supports growth of the organism being tested. This challenge testing approach requires homogeneous distribution of the microbes which is crucial regardless of the food being solid, liquid, semiliquid, or powder. Food structure can be described as the spatial arrangement of the structural elements of food products and their interactions [1, 7, 22]. Although food structural properties can be analyzed at different dimensions, including molecular, nanoscale, microscale, and macroscale level (Fig. 1.), the microscale level is most relevant when studying microbial behavior [1]. Even for motile organisms, the homogeneous distribution of cells in each microscopic site is particularly constrained in gelled/solid food matrices [1, 6, 23] which can be a major cause of discrepancies in predictive microbiology when liquid models are used.

Microbes in food matrices are found in one of three growth morphologies: planktonic cells, submerged colonies and surface colonies, or a combination of these [24]. Predictive growth models are routinely developed based on data from laboratory media and are applied to the broad range of structured food matrices such as aqueous gels, emulsions, gelled emulsions, and others. Table 2 illustrates broad food matrices and their structural effect on microbial growth. Basically, microbial growth primarily occurs in the

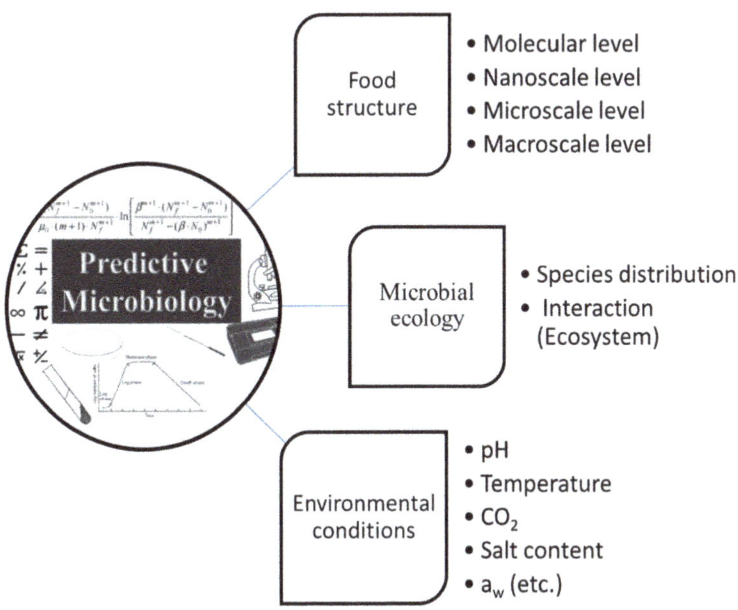

Fig. 1 Schematic diagram of predictive microbiology

Table 2
Food matrices and their structural effect on microbe's growth

Matrix	Structure	Example food	Influence description	Reference
Liquid	Uniform structure	Soups, juices	Little or no effect except gravity	[5]
Semisolid	Gel	Yogurt, jellies, cheeses	Localized changes in nutrient availability and accumulation of metabolites	[23, 26]
	Oil-in-water emulation	Dairy cream, milk, salad cream, mayonnaise	Oil phase in a matrix limits bacterial motility	[22, 27]
	Water-in-oil emulation	Butter, margarine, low-fat spread	A microorganism present in a droplet would be restricted in its growth either by space limitation or by nutrient depletion	[4, 23, 28]
	Gelled emulation	Whole-milk cheese, sausage	Contain gums and thus restrict mobilization of microbes	[5, 6, 23, 29]
Tissue	Surface	Meat, vegetables	Similar to a gel but colony growth constraint diffusion at a surface is greater than within a gel	[30, 31]
Powder	Amorphous, crystalline, or mixed structure	Milk powder, chili powder	The structural heterogeneity and uneven a_w may allow localized microbial growth	[26]

aqueous phase of food. Planktonic (free-swimming) growth morphology is observed in liquid food matrix (liquid food). This allows motile bacteria to migrate to their preferred location, i.e., toward nutrients and away from their metabolites [23]. Hence, planktonic growth is commonly assumed to have the highest growth rate, followed by submerged colonies and slowest as surface colonies [12]. For semisolid food formed by gelling agents (gum, gelatine, etc.), microbial cells are trapped within the gelled regions and restricted to grow as submerged colonies in the structured aqueous phase of the gel [9]. The localized changes in nutrient availability and accumulation of metabolites within such an aqueous phase of the gel limit the growth. Furthermore, when there is a higher fat content in the matrix, growth can be restricted even further.

Microbial growth on the surfaces of the food matrix is colonial and concentric [24]. Nutrient replenishment comes from the bottom or periphery of the colony, and cells in the center quickly become starved which suppresses the growth of the microbes. In contrast, powdered food matrices, such as milk powder, flours, and herbs, do not support microbial growth due to their low a_w (*see* detail in Subheading 1.3). However, microorganisms do survive in powdered foods for prolonged periods, e.g., *Salmonella*, *Staphylococcus aureus*, *Bacillus cereus*, and *Cronobacter sakazakii* in milk powder [25].

1.3 Water Activity (a_w)

Water activity (a_w) is a physiochemical parameter, which is a measure of the energy status of the water in a matrix. It determines the water available in the product, meaning the "free" water that is not chemically bound to other molecules like proteins, carbohydrates, or sugars. Water activity is defined as the ratio between the vapor pressure of the water in food (ρ) and the vapor pressure of pure distilled water (ρ_0) under identical conditions, such as temperature and pressure ($a_w = \rho / \rho_0$). Typically, as the temperature increases, a_w also increases. Water activity is expressed as a decimal, with a scale running from 0.0 to 1.0, where 0.0 indicates an absence of water and 1.0 represents pure water. It's important to note that water activity is not the same as water content. Water content is simply the amount of water in the product, both "free" and "bound." For example, products such as jam and honey have a high moisture content, but these products are very high in sugar which binds the free water, thus leaving very little water available for the microbes. Similarly, brine, a saline solution used to cure meats, is very high in salt, which exerts osmotic pressure on microbes, thereby reducing the turgor pressure and viability of the cells.

While all microorganisms require moisture for growth, the exact amount of water needed varies between bacterial species, molds, and yeasts (Table 3). Each microorganism copes with osmotic stress in a different way and has a different limit of water activity, below which it will not grow. Most microbes flourish well in food with an a_w of >0.98, but most spoilage and pathogenic molds and bacteria can only tolerate a minimum level of 0.70–0.90, and there is no microbial proliferation below 0.60 [32, 33]. These limits are usually determined in adjusted laboratory media and do not account for local variation in a_w in real food matrices.

1.4 Growth of Microbes in Acidified and Acidic Food

Microbes respond to acid stress by regulating intracellular pH in response to changes in the environment by modifying the lipid composition of the cytoplasmic membrane to reduce the permeability of protons [34] or by refluxing protons by H^+-ATPase transporters [35] or using several metabolic mechanisms such as the glycolytic pathway [36]. To limit microbial proliferation and assure food organoleptic stability and safety, low pH or acidification is commonly employed in food preservation methods such as pickling or fermentation, and the pH values of various foodstuffs are available elsewhere [37, 38]. In general, most pathogen in food are inhibited at pH levels below 4.2, whereas lactic acid bacteria and several species of yeasts and molds grow at pH levels considerably below this [37, 38]. pH represents the hydrogen ion concentration in a sample matrix (aqueous solution), and the higher the hydrogen ion in the food matrix, the lower the pH (Eq. 1). Strong acids have low acid dissociation constant (p*K*a) reflecting a stronger ability to donate protons. As a consequence, strong acids dissociate entirely in food matrices, having less of an influence on the microbes there.

Table 3
Growth of microbes at range of a_w and food

Range of a_w	Example food	Inhibited organisms
1.00–0.95	Canned fruits, vegetables, meat, fish, milk, and beverages	*Pseudomonas*, *Escherichia*, *Proteus*, *Shigella*, *Klebsiella*, *Bacillus*, *Clostridium perfringens*, *C. botulinum* E, G, some yeasts
0.95–0.91	Cheese, cured meat (ham), bread, tortillas	*Salmonella*, *Vibrio parahaemolyticus*, *Clostridium botulinum* A, B, *Listeria monocytogenes*, *Bacillus cereus*
0.91–0.87	Fermented sausage (salami), sponge cakes, dry cheeses, margarine	*Staphylococcus aureus* (aerobic), many yeasts (*Candida*, *Torulopsis*, *Hansenula*), *Micrococcus*
0.87–0.80	Fruit juice concentrates, sweetened condensed milk, syrups, jams, jellies	Most molds (*mycotoxigenic penicillia*), *Staphylococcus aureus*, most *Saccharomyces* (bailii) spp., *Debaryomyces*
0.80–0.75	Marmalade, marzipan, glacé fruits, beef jerky	Most halophilic bacteria, mycotoxigenic aspergilli
0.75–0.65	Molasses, raw cane sugar, some dried fruits, nuts, snack bars, snack cakes, dry milk	Xerophilic molds (*Aspergillus chevalieri*, *A. candidus*, *Wallemia sebi*), *Saccharomyces bisporus*
0.65–0.61	Dried fruits	Osmophilic yeasts (*Saccharomyces rouxii*), a few molds (*Aspergillus echinulatus*, *Monascus bisporus*)
< 0.61	Dry pasta, rice, chocolate	No microbial proliferation

Adapted from Beuchat [32]. IFT Continuing Education Committee, June 14–15, Anaheim, CA

For example, hydrochloric acid with molecular weight of 36.46 g/mol and p*K*a value <1 tends to dissociate completely in aqueous solutions:

$$\mathrm{pH} = -\log 10\,[\mathrm{H}^+] \tag{1}$$

$$(\text{Henderson} - \text{Hasselbalch})\ \mathrm{pH} = \mathrm{p}Ka + \log_{10}\left(\frac{[\text{Base}]}{[\text{Acid}]}\right) \tag{2}$$

where *K*a is the dissociation constant of the weak acid and [Acid] and [Base] are the molar concentration of the weak acid and its conjugate base.

Acetic, benzoic, citric, lactic, sorbic, and propionic acids are widely used in food preservation and generally regarded as safe by authorities (*see* Table 4 for detail). The importance of the weak acid or more importantly the undissociated weak acid is that, unlike charged anions, their molecular structure is lipid-soluble and they can permeate cells by simple diffusion through the plasma membrane into the cytoplasm at low pH [39]. The partial dissociation characteristic of weak acid in an aqueous solution causes membrane disruption [40–42], inhibition of essential metabolic reactions

Table 4
Commonly used organic acid in food industry

Weak acid	p*Ka*	E-number	Maximum level	Application
Acetic acid	4.74	E260	*Quantum satis*	Mayonnaise, salad dressings, pickles, chutney, ketchups, cheeses, and baked products
Benzoic acid	4.19	E210	1000 mg/kg	Only cooked shrimps in brine
Citric acid	2.79	E330	*Quantum satis*	Wines, ciders, soft drinks, syrups, ice creams, andprocessed cheese
Lactic acid	3.79	E270	*Quantum satis*	Salad dressing, olives, cheese, frozen desserts, carbonated drinks
Sorbic acid	4.75	E200	1500 mg/kg	Liquid egg, cooked seafood, soft drinks, wine
Propionic acid	4.87	E280	1000 mg/Kg- *quantum satis*	Bread, tortillas, cheeses, cakes

$$CH_3CH_2COOH \rightleftharpoons CH_3CH_2COO^- + H^+ \quad \text{Equation 4}$$

Weak acid (propionic) | propionate anions | hydrogen ion

Fig. 2 Acid dissociation reaction of propionic acid

[43], perturbation of intracellular pH homeostasis [44, 45], and the accumulation of toxic anions to the microbes [46, 47]. Basically, the dissociated and undissociated forms of weak acids exist in solution in a pH-dependent equilibrium (Fig. 2). Lactic acid, for example, has a p*Ka* of 3.86 which means that at a pH of 3.86, 50% of the lactic acid is dissociated and 50% is undissociated. While the mechanism of action on microbe's inhibition is similar in general, there may be differences in the efficacy of undissociated acids against microbes. Sorbic and acetic acid have similar p*Ka* values (4.76), but more acetic acid is required to have equal growth-inhibitory effects [48, 49].

As with other intrinsic food properties, the effect of organic acids on microbial growth rates can be efficiently modeled in laboratory media. However, there can be significant discrepancies between the behavior of weak acids in an aqueous laboratory media compared to a liquid or semisolid food matrices [50].

1.5 Alignment with Challenge Tests

The traditional method to determine the susceptibility of a food matrix to microbial growth is the performance of a challenge test. There are a number of critical assessments on the planning and execution of challenge trials which address the most fundamental requirement of accurately mimicking the conditions that the challenge organisms would experience in the actual food [52–54]. In a food matrix, the test organisms (or representative surrogates) must be the organisms of concern in the test matrix, the challenge inoculum level must be the level likely to contaminate, and the appropriate duration (shelf life) and types of analysis (presence/absence, ability of growth or growth potential) must reflect what would likely happen in real-life circumstances. The European Union Reference Laboratory (EURL) guidelines define the use of three batches of the same food matrix for a challenge test and use of well-characterized EURL strains [51]. In general, strain cocktails are recommended to account for strain variability [51, 52]. Additional details on parameters to consider when designing challenge studies were published by the National Advisory Committee on Microbiological Criteria for Foods [53] and the European Food Safety Authority [54].

2 Materials

Accurate measurement of the intrinsic food properties is critical. Therefore, careful selection of instrumentation and sampling strategies is required.

2.1 Determination of a_w in Food

There are four basic water activity instruments: resistive electrolytic hygrometers (REH), capacitance sensors, chilled mirror dew point hygrometers, and tunable diode laser (TDL). As each instrument has its advantages and disadvantages, the choice of instrument typically depends on the type of matrix to be analyzed. The typical instrument consists of a sealed, temperature- and pressure-controlled container where a sample is placed. Free water from the sample escapes into the air until a condition of equilibrium is met.

Resistive Electrolytic Hygrometers (REH) This instrument is fitted with a liquid electrolyte held in between of two small glass rods by the capillary force. As the water from the sample escapes, the resistance of the electrolyte is affected. The resistance is directly proportional to relative air humidity and to a_w of the sample, and this relationship can be checked by either a verification or calibration using salt-water mixtures. This type of instrument is affected by most volatile chemicals, but special protection filters can be used to absorb the volatile compounds, but routine calibration is required.

Capacitance Sensors Capacitance sensors are least accurate and require regular calibration checks, but they are not affected by most volatile chemicals. The instrument is fitted with two charged plates separated by a polymer membrane dielectric, and capacitance is controlled. When the polymer membrane absorbs water from the sample, its ability to hold a charge increases; therefore, capacitance measure also increases. The capacitance is approximately proportional to the a_w as determined by a sensor specific calibration.

Chilled Mirror Dew Point Hygrometers These are accurate instruments for higher water activities and require no calibration. A polished metal mirror is placed over a closed sample container and is cooled until the dew forms and condensates on the mirror. The temperature is measured by an optical sensor fitted in the instrument. The dew point temperature is directly related to the vapor pressure of the air, so the a_w can be calculated.

Tunable Diode Laser (TDL) This is commonly used to analyze products with volatile compounds (such as pure alcohol, propylene glycol). TDL features a finely tuned laser sensor and provides precise results in a much faster time than in the other instruments (< 5 min).

2.2 Determination of pH in Food

To measure the pH of food matrices, numerous commercial potentiometric pH meters are available from portable pen like device to table top apparatus. In general, there are two types of probes: surface probes, which are used for solid foods, and liquid probes, which are used for liquid foods. Semisolid and solid food sample can be prepared as one in ten diluted homogenates with sterile distilled water before taking pH measurement [50]. The colorimetric pH strips ("MColorpHast™" pH 5.2–7.2, Merck KGaA, Darmstadt, Germany) have been used to measure pH of the muscle meat by inserting test strip into the incision, and digital camera was used to analyze the image [51]. In addition, a colorimetric pH meter can also be used to measure pH, but most foods produce chemical interactions that destroy color indicators through oxidation or reduction, and the FDA only allows this approach to be used on foods with a pH of less than 4.

3 Method

3.1 Inoculum Preparation

1. Inoculate a single colony from the agar plate into 10 mL sterile broth, and incubate at optimal temperature overnight (*see* **Notes 1–5**).
2. Washing the strain: Centrifuge 1 mL of ON culture at 13,000 rpm for 5 min, and remove supernatant and resuspend in 1 mL 0.85% NaCl, and vortex to resuspend. Repeat the procedure twice to avoid transfer of any media components to the test matrix that could affect microbial growth (*see* **Note 5**).
3. For preparation of strain cocktails (optional), mix equal amounts (1 mL) of each strain standardized by optical density, for example, 0.1 ± 0.02 OD_{600}.

3.2 Preparation of Test Matrix

1. Sterility confirmation: Dilute 1 g of each food sample in 9 mL of physiological water. Vortex thoroughly and spread 100 μL of aliquots on plate count agar. Incubate at 30 °C for 48 h, and inspect the plates for signs of product contamination.
2. Water activity adjustment: Table 5 shows an example of a_w adjustment in laboratory broth (BHI) with salt, and similar adjustments are possible with liquid and semiliquid food matrices (*see* **Note 6**). The appropriate amount of sodium chloride (NaCl) (BioXtra, ≥99.5% (AT)) (Sigma Aldrich, Reference, S7653-1KG) was added, and an AQUALAB 4TEV water activity meter (METER Group, Munich, Germany) was used to generate data shown in Table 5.
3. Acidification of test matrices: Where required, the pH of liquid and semiliquid food products should be adjusted using 1 M hydrochloric acid (HCl) ACS reagent, 37% (Sigma Aldrich, Reference, 320,331–500 mL), or 1 M sodium hydroxide (NaOH) (≥98%, pellets (anhydrous), Sigma Aldrich, Reference, S5881-500G). Table 6 shows an example for calculating undissociated organic acid (acetic and propionic) concentration in food matrices adjusted to pH 4.7, 5, and 5.3 with HCl. The acid/salt buffer volumes required to reach a desired concentration of undissociated acid were calculated using the Henderson-Hasselbalch equation (Eq. 2). Buffers were prepared at the required pH, so, for example, to prepare 1 L of 100 mM sodium acetate acid buffer at pH 5.3 (undissociated acid concentration 220 mM), add 13.20 mL acetic acid 99–100% (Sigma Aldrich, Reference, 8.18755) and 63.99 g sodium acetate (ACS reagent, ≥99.0%, Sigma Aldrich, Reference, 241,245-500G) in ~990 mL pure water. pH was measuring pH meter (Mettler Toledo, Ohio, USA).

Table 5
Water activity adjustment by addition of NaCl in a laboratory broth (BHI)

Sample	NaCl (g)	T (°C)	a_w value
BHI	0.00	22.50	0.990
	0.25	23.20	0.987
	0.50	23.00	0.980
	0.75	23.50	0.972
	1.00	23.40	0.968
	1.25	23.40	0.962
	1.50	23.80	0.952
	1.75	23.20	0.946
	2.00	23.20	0.939
	2.25	22.80	0.936
	2.50	23.30	0.919

Table 6
Example of acetic acid and propionic acid adjustment table

Buffer	pH	[AH] in buffer (mM)	[AH] in matrix (mM)	Matrix vol. (mL)	Buffer V to add (mL)
Acetate	4.7	52.9	1	20	0.378
Acetate	5	36.0	1	20	0.556
Acetate	5.3	22.0	1	20	0.910
Propionate	4.7	60.2	1	20	0.332
Propionate	5	43.1	1	20	0.464
Propionate	5.3	27.5	1	20	0.726
Acetate	4.7	52.9	2	20	0.757
Acetate	5	36.0	2	20	1.111
Acetate	5.3	22.0	2	20	1.819
Propionate	4.7	60.2	2	20	0.664
Propionate	5	43.1	2	20	0.927
Propionate	5.3	27.5	2	20	1.452

3.3 Growth Experiment

1. Add the appropriate amount of the inoculum to the matrix to achieve the required starting concentration. For example, 100 μL of a 10^5 CFU/mL culture is inoculated into 9.9 mL of test matrix to give a starting inoculum level of 10^3 in the test matrix in a 15 mL test tube (*see* **Note 1**). Inoculum volume should be minimized to avoid alteration of food matrix formulation.
2. Mix inoculated cells through the food by vortexing liquid matrices or by stirring with a sterile inoculation loop for the semisolid to try and ensure even distribution of the test microbes in the matrix (*see* **Note 7**). For other matrices, *see* **Notes 8** to **10**.
3. Enumerate immediately for accurate starting (T_0) bacterial counts. Serially dilute 100 μL sample from each sample and spread on the appropriate agar, or microdrop method can be used to minimize consumables (*see* **Note 11**). Incubate the plates at the appropriate temperature, and record the colony counts after 24–72 h depending on the microbe.
4. Incubate the test units at appropriate temperature according to experimental design, and enumerate the test samples at least five to seven times starting from T_0. *See* **Note 12** for estimating sampling times.
5. Fit the data in the ComBase online tool, or downloadable Excel worksheet freely available at https://browser.combase.cc/DMFit.aspx to calculate the growth rate.

4 Notes

1. Prior to experiment setup, the ON CFU/mL concentrations of all test strains should be determined by standard plating of serial diluted cell suspensions on suitable agar plates. This will help achieve correct starting inoculum levels.
2. For each growth experiment, at least three biological replicates are recommended, i.e., three repeats of the experiment from different colonies on the agar plate. For quality control and to monitor variables of the process, growth experiments should also be completed on different days. Uninoculated units are employed as negative controls to detect levels of natural contamination from background microflora.
3. All the organisms under study should be stocked in 40% glycerol (stored at −20 °C short term and − 80 °C long term) to prevent changes in morphological, physiological, and genetic characteristic to ensure reproducible results. Storage of strains for growth experiments on agar plate at 4 °C is recommended for not more than 4 weeks.
4. Subculturing is required to obtain microbes in an unstressed state and to help in consistency of inoculum concentration. However, repeated subculturing is not recommended since it can cause contamination, viability loss, and genotypic/phenotypic alterations.
5. It should be noted that the state of the cells cultured ON in laboratory media is unlikely to mimic starting inoculum in food environments which may be stressed or already preadapted.
6. To adjust a_w or pH in semisolid food such as puree or sauces, add the estimated amount of filter-sterilized salt, acid, or NaOH. Mix thoroughly, and remove several (four to five) randomly selected samples to measure the a_w or pH separately and discard the portion.
7. Mechanical mixing such as with an inoculating loop is suitable for small sample volumes only, and distribution should be checked by comparing starting inoculum CFU/mL in several randomly selected samples from the test matrix. Where inoculation distribution is expected to be uneven, a larger number of samples at each sampling interval should be examined. Mixing should be consistent across all experimental replicates, e.g., 30 s vortexing or 60 s stirring with inoculating loop.
8. It is impossible to distribute evenly in tissue foods such as fruit slices, cheese, or processed meat because growth occurs primarily on the surfaces of such matrices. The appropriate concentration of inoculum in the smallest volume should be applied aseptically to the test unit prior to the experiment [56], and inoculating individual portions (e.g., 3 mm-thick

slice) is recommended. After the inoculation in solid and powder matrices, a set holding period is required to allow the inoculum to attach to the product which such be conducted in a laminar airflow cabinet.

9. To recover cells from solid or tissue foods, test units are put into sterile stomacher bags at each sampling point, diluted 1:5 with physiological water, and homogenized for 30 s in a stomacher followed by serial dilution and plating on agar plates.

10. For foods that need to be tested in-pack, ideally the matrix is inoculated and then packaged under modified atmosphere or aseptic conditions. However, inoculation is possible through the packaging with fixed self-sealing rubber or silicon septum with a sterile needle and syringe.

11. Selective laboratory media can be used although medium selection should consider the recovery of stressed cells. It is generally accepted that reasonably accurate counts are between 25 and 300 colonies for the spread plate (i.e., 100 μL spread) method [54]. To minimize the cost of consumables and reduce time, multiple dilution of 10 μL diluted sample can be plated as a drop on one agar plate instead, and it is widely used. The acceptable or countable range of the colony for the drop plate method is between 3 and 30 [55].

12. To generate reliable and accurate growth rate data, a suitable spread of sampling points is essential. Suitable sampling time points can be estimated using the ComBase predictor where test organism data is available. Alternatively, literature searches can help establish suitable sampling points. However, it is often necessary to preform preliminary experiments to identify suitable time point intervals for growth rate modeling. This is particularly important when investigating hurdle efficacy in food matrices where growth rates are often significantly reduced and/or subtly different. Finally, the risk of contamination caused by frequently resampling can be minimized by using multiple batches of discrete test units at each sampling interval.

References

1. Verheyen D, Van Impe JFM (2021) The inclusion of the food microstructural influence in predictive microbiology: state-of-the-art. Foods 10(9):2119. https://doi.org/10.3390/foods10092119
2. Baka M, Verheyen D, Cornette N, Vercruyssen S, Van Impe JF (2017) *Salmonella typhimurium* and *Staphylococcus aureus* dynamics in/on variable (micro)structures of fish-based model Systems at Suboptimal Temperatures. Int J Food Microbiol 240:32–39. https://doi.org/10.1016/j.ijfoodmicro.2016.08.004
3. Nyhan, L.; Johnson, N.; Begley, M.; Callanan, M. An evaluation of lux technology as an alternative methodology to determine growth rates of *listeria* in laboratory media and complex food matrices. Int J Food Microbiol, 2020, 317 (February 2019), 108442. https://doi.org/10.1016/j.ijfoodmicro.2019.108442
4. Robins MM, Wilson PDG (1994) Food structure and microbial growth. Trends Food Sci

Technol 5:289–293. https://doi.org/10.1016/0924-2244(94)90137-6

5. Verheyen D, Baka M, Glorieux S, Duquenne B, Fraeye I, Skåra T, Van Impe JF (2018) Development of fish-based model systems with various microstructures. Food Res Int 106:1069–1076. https://doi.org/10.1016/j.foodres.2017.12.047
6. Verheyen D, Bolívar A, Pérez-Rodríguez F, Baka M, Skåra T, Van Impe JF (2018) Effect of food microstructure on growth dynamics of *listeria monocytogenes* in fish-based model systems. Int J Food Microbiol 283 (February):7–13. https://doi.org/10.1016/j.ijfoodmicro.2018.05.032
7. Aguilera JM (2005) Why food microstructure? J Food Eng 67:3–11. https://doi.org/10.1016/j.jfoodeng.2004.05.050
8. Baranyi J, Roberts TA (1994) A dynamic approach to predicting bacterial growth in food. Int J Food Microbiol 23(3–4):277–294. https://doi.org/10.1016/0168-1605(94)90157-0
9. Skandamis PN, Jeanson S (2015) Colonial vs. planktonic type of growth: Mathematical modeling of microbial dynamics on surfaces and in liquid, semi-liquid and solid foods. Front Microbiol 0(OCT):1178. https://doi.org/10.3389/FMICB.2015.01178
10. Baka M, Noriega E, Van Langendonck K, Van Impe JF (2016) Influence of food intrinsic complexity on *listeria monocytogenes* growth in/on vacuum-packed model Systems at Suboptimal Temperatures. Int J Food Microbiol 235:17–27. https://doi.org/10.1016/j.ijfoodmicro.2016.06.029
11. O'Toole DK (1983) Methods for the direct and indirect assessment of the bacterial content of Milk. J Appl Bacteriol 55(2):187–201. https://doi.org/10.1111/J.1365-2672.1983.TB01315.X
12. Kogure K (1979) A tentative direct microscopic method for counting living marine bacteria rapid screening of enteric pathogens from the environment view project development of marine ecosystem evaluation methods in the high throughput sequencing era view project. Artic Can J Microbiol
13. Sysel AM, Dunphy MJ, Bauer JA Antimicrobial properties of Diethylamine NONOate, a nitric oxide donor, against Escherichia Coli: A pilot study. J Antibiot 2020 74(4):260–265. https://doi.org/10.1038/s41429-020-00397-9
14. Diez-Gonzalez F (2014) *Total viable counts: specific techniques*, vol 3, 2nd edn. Elsevier. https://doi.org/10.1016/B978-0-12-384730-0.00332-3
15. Sanders ME, Merenstein DJ, Ouwehand AC, Reid G, Salminen S, Cabana MD, Paraskevakos G, Leyer G (2016) Probiotic use in at-risk populations. J Am Pharm Assoc 56(6):680–686. https://doi.org/10.1016/j.japh.2016.07.001
16. Wilkinson MG (2018) Flow cytometry as a potential method of measuring bacterial viability in probiotic products: A review. Trends Food Sci Technol 78(May):1–10. https://doi.org/10.1016/j.tifs.2018.05.006
17. International Organization for Standardization. Milk and milk products — Starter cultures, probiotics and fermented products — Quantification of lactic acid bacteria by flow cytometry https://www.iso.org/obp/ui/#iso:std:iso:19344:ed-1:v1:en. Accessed 29 Nov 2021
18. Huhtanen CN, Jones CO (1989) A procedure for the direct microscopic count of bacteria in non-fat dry milk. J Food Prot 52(6):404–407. https://doi.org/10.4315/0362-028X-52.6.404
19. Sharma S, Kaur S, Kumar A, Sharma N, Kapoor P, Chunduri V, Kumari A, Garg M (2020) Diagnostic tools for food safety. In: Raj T, Rupesh S, Humira D (eds) Advances in Agri-food biotechnology. Singapore, Springer Nature, pp 333–365
20. Cao R, Wan Q, Tan L, Xu X, Wu G, Wang J, Xu H, Huang T, Wen G (2021) Evaluation of the vital viability and their application in fungal spores' disinfection with flow cytometry. Chemosphere 269:1–12. https://doi.org/10.1016/j.chemosphere.2020.128700
21. Dumont MG, Murrell JC (2005) Stable isotope probing - linking microbial identify to function. Nat Rev Microbiol 3(6):499–504. https://doi.org/10.1038/nrmicro1162
22. Heertje I (2014) Structure and function of food products: A review. Food Struct 1(1):3–23. https://doi.org/10.1016/j.foostr.2013.06.001
23. Wilson PDG, Brocklehurst TF, Arino S, Thuault D, Jakobsen M, Lange M, Farkas J, Wimpenny JWT, Van Impe JF (2002) Modelling microbial growth in structured foods: towards a unified approach. Int J Food Microbiol 73:275–289
24. Wimpenny JWT (1992) Microbial systems. Adv Microb Ecol 12:469–521. https://doi.org/10.1021/cen-v065n013.p024
25. Lang E, Sant'Ana AS (2021) Microbial contaminants in powdered infant formula: what is the impact of spray-drying on microbial

inactivation? Curr Opin Food Sci 42:195–202. https://doi.org/10.1016/j.cofs.2021.06.007

26. Jongenburger I, Reij MW, Boer EPJ, Zwietering MH, Gorris LGM (2012) Modelling homogeneous and heterogeneous microbial contaminations in a powdered food product. Int J Food Microbiol 157(1):35–44. https://doi.org/10.1016/j.ijfoodmicro.2012.04.009
27. Parker ML, Brocklehurst TF, Gunning PA, Coleman HP, Robins MM (1995) Growth of food-borne pathogenic bacteria in oil-in-water emulsions: I—methods for investigating the form of growth. J Appl Bacteriol 78(6): 601–608. https://doi.org/10.1111/j.1365-2672.1995.tb03105.x
28. Tsironi T, Houhoula D, Taoukis P (2020) Hurdle technology for fish preservation. Aqua Fish 5(2):65–71. https://doi.org/10.1016/J.AAF.2020.02.001
29. Costello KM, Gutierrez-Merino J, Bussemaker M, Ramaioli M, Baka M, Van Impe JF, Velliou EG (2018) Modelling the microbial dynamics and antimicrobial resistance development of *listeria* in viscoelastic food model Systems of Various Structural Complexities. Int J Food Microbiol 286:15–30. https://doi.org/10.1016/j.ijfoodmicro.2018.07.011
30. Brocklehurst TF, Mitchell GA, Smith AC (1997) A model experimental gel-surface for the growth of bacteria on foods. Food Microbiol 14(4):303–311. https://doi.org/10.1006/FMIC.1997.0098
31. Brocklehurst TF, Mitchell GA, Smith AC (1997) A model experimental gel surface for the growth of bacteria on foods. Food Microbiol 14:303–311
32. Tapia MS, Alzamora SM, Chirife J (2020) Effects of water activity (a_w) on microbial stability: as a hurdle in food preservation. In: Water activity in foods: fundamentals and applications, Chicago, pp 239–271. https://doi.org/10.1002/9780470376454.ch10
33. Beuchat LR (2002) Water Activity and Microbial Stability. In: *Fundamentals of water activity*, pp 239–272
34. Shabala L, Ross T (2008) Cyclopropane fatty acids improve Escherichia Coli survival in acidified minimal media by reducing membrane permeability to H+ and enhanced ability to extrude H+. Res Microbiol 159(6):458–461. https://doi.org/10.1016/j.resmic.2008.04.011
35. Sun Y (2016) F1 F0-ATPase functions under markedly acidic conditions in bacteria. Springer International Publishing, Cham, pp 459–468. https://doi.org/10.1007/978-3-319-24780-9_22
36. Guan N, Li J, Shin HD, Du G, Chen J, Liu L (2016) Metabolic engineering of acid resistance elements to improve acid resistance and propionic acid production of *Propionibacterium jensenii*. Biotechnol Bioeng 113(6): 1294–1304. https://doi.org/10.1002/bit.25902
37. Institute of Food Technologists the Food Drug Administration (2001) Evaluation and definition of potentially hazardous foods. Food Drug Adm:21–32
38. Rahman MS (2007) pH in food preservation. In: Handbook of food preservation. CRC Press, Boca Raton, pp 323–332
39. Stratford M, Nebe-von-Caron G, Steels H, Novodvorska M, Ueckert J, Archer DB (2013) Weak-acid preservatives: pH and proton movements in the yeast *Saccharomyces cerevisiae*. Int J Food Microbiol 161(3):164–171. https://doi.org/10.1016/j.ijfoodmicro.2012.12.013
40. Bracey D, Holyoak CD, Coote PJ (1998) Comparison of the inhibitory effect of Sorbic acid and amphotericin B on *Saccharomyces cerevisiae*: is growth inhibition dependent on reduced intracellular pH? J Appl Microbiol 85(6):1056–1066. https://doi.org/10.1111/j.1365-2672.1998.tb05271.x
41. Freese E, Sheu CW, Galliers E (1973) Function of lipophilic acids as antimicrobial food additives. Nature 241(5388):321–325. https://doi.org/10.1038/241321a0
42. Stratford M, Anslow PA (1998) Evidence that sorbic acid does not inhibit yeast as a classic "Weak Acid Preservative". Lett Appl Microbiol 27(4):203–206. https://doi.org/10.1046/j.1472-765X.1998.00424.x
43. Krebs HA, Wiggins D, Stubbs M, Sols A, Bedoya F (1983) Studies on the mechanism of the antifungal action of benzoate. Biochemist 214(3):657–663. https://doi.org/10.1111/j.1469-8137.1979.tb07471.x
44. Cole MB, Keenan MHJ (1987) Effects of weak acids and external pH on the intracellular pH of *Zygosaccharomyces bailii*, and its implications in weak-acid resistance. Yeast 3(1):23–32. https://doi.org/10.1002/YEA.320030105
45. Salmond CV, Kroll RG, Booth IR (1984) The effect of food preservatives on pH homeostasis in *Escherichia coli*. J Gen Microbiol 130(11): 2845–2850. https://doi.org/10.1099/00221287-130-11-2845
46. Eklund H, Horjales E, Joernvall H, Braenden C-I, Jeffery J (1985) Molecular aspects of

functional differences between alcohol and sorbitol dehydrogenases. Biochemistry 24(27): 8005–8012. https://doi.org/10.1021/BI00348A025

47. Roe AJ, McLaggan D, Davidson I, O'Byrne C, Booth IR (1998) Perturbation of anion balance during inhibition of growth of *Escherichia coli* by weak acids. J Bacteriol 180(4):767
48. Ter Beek A, Wijman JGE, Zakrzewska A, Orij R, Smits GJ, Brul S (2015) Comparative physiological and transcriptional analysis of weak organic acid stress in *Bacillus subtilis*. Food Microbiol 45(PA):71–82. https://doi.org/10.1016/j.fm.2014.02.013
49. Heavin SB, Brennan OM, Morrissey JP, O'Byrne CP (2009) Inhibition of *listeria monocytogenes* by acetate, benzoate and sorbate: weak acid tolerance is not influenced by the glutamate decarboxylase system. Lett Appl Microbiol 49(2):179–185. https://doi.org/10.1111/j.1472-765X.2009.02634.x
50. Nyhan L, Begley M, Mutel A, Qu Y, Johnson N, Callanan M (2018) Predicting the combinatorial effects of water activity, pH and organic acids on *listeria* growth in media and complex food matrices. Food Microbiol 74: 75–85. https://doi.org/10.1016/j.fm.2018.03.002
51. Bergis H, Bonanno L, Asséré A, Lombard B (2021) EURL lm TECHNICAL GUIDANCE DOCUMENT on challenge tests and durability studies for assessing shelf-life of ready-to-eat foods related to *listeria monocytogenes*. EURL Lm 4:60
52. Myintzaw, P.; Pennone, V.; McAuliffe, O.; Begley, M.; Callanan, M. Correlation of organic acid tolerance and genotypic characteristics of *listeria monocytogenes* food and clinical isolates. Food Microbiol, 2022, 104 (July 2021). https://doi.org/10.1016/j.fm.2022.104004
53. (2010) National Advisory Committee on microbiological criteria for foods (NACMCF), Parameters for determining inoculated pack/challenge. J Food Prot 73(1):140–202
54. Andreoletti O, Budka H, Buncic S, Colin P, Collins JD, De A, Griffin J, Havelaar A, Hope J, Klein G et al (2008) Request for updating the former SCVPH opinion on *listeria monocytogenes* risk related to ready-to-eat foods and scientific advice on different levels of *listeria monocytogenes* in ready-to-eat foods and the related risk for human illness - scientific Opinio. EFSA J 6(1):1–42. https://doi.org/10.2903/j.efsa.2008.599
55. Thomas P, Sekhar AC, Upreti R, Mujawar MM, Pasha SS (2015) Optimization of single plate-serial dilution spotting (SP-SDS) with sample anchoring as an assured method for bacterial and yeast Cfu enumeration and single Colony isolation from diverse samples. Biotechnol Rep 8:45–55. https://doi.org/10.1016/j.btre.2015.08.003
56. Herigstad B, Hamilton M, Heersink J (2001) How to optimize the drop plate method for enumerating bacteria. J Microbiol Methods 44(2):121–129. https://doi.org/10.1016/S0167-7012(00)00241-4

Chapter 4

Primary Predictive Models of Microbial Growth

Isabella Bassoto Xavier, Jean Carlos Correia Peres Costa, and Verônica Ortiz Alvarenga

Abstract

Foodborne illness occurs when food or beverages are contaminated with pathogenic bacteria, chemicals, or toxins. Understanding microbial physiology and predicting microbial behavior, such as their growth, are fundamental for food safety. The development of microbial growth modeling stemmed from the criticism that food quality control based on the challenge tests of the final products is expensive, labor-intensive, and time-consuming. This chapter aims to develop a practical protocol on how to apply growth models to microbial behavior.

Key words Microbial growth models, Foodborne illness, Foodborne pathogens, Food spoilage, Primary models

1 Introduction

Foodborne illness is caused by consuming contaminated food by pathogen organisms, chemicals, or other agents. The Centers for Disease Control and Prevention (CDC) estimates that there are about 48 million cases of foodborne illness each year, which results in an estimated 128,000 hospitalizations and 3000 deaths [1, 2]. The most common foodborne pathogens are *Escherichia coli*, *Salmonella* spp., *Listeria monocytogenes*, and hepatitis A virus [1]. Therefore, it is essential to examine foods for the presence of pathogens and their growth to increase food safety. Conventional detection methods, such as viable plate counts, are commonly employed to test food product safety [2].

Microbial growth is a fundamental process in microbiology and refers to the increase in number of a population of microorganisms. Several studies approach predictive models to explain microbial behavior in different environments [3, 4]. Predictive microbiology is an area of study that integrates knowledge of food microbiology with important mathematical and modeling concepts to develop

Verônica Ortiz Alvarenga (ed.), *Basic Protocols in Predictive Food Microbiology*, Methods and Protocols in Food Science,
https://doi.org/10.1007/978-1-0716-3413-4_4,

models capable of describing and predicting the behavior of microorganisms under specified environmental conditions [5–7]. For food industries, application of this knowledge can provide important benefits to ensure food safety and quality, from farm to fork.

The development of growth models emerged in the last 25 years as a tool to help in the difficulties of traditional methods of microbial counting in foods, based on challenge tests, and also in decision-making, due to the rapid responses to specific questions [5–7]. Several mathematical models have been applied as growth equations to describe microbial growth. These equations differ from each other in terms of "ease of use" and the number of parameters each equation has. Microbial growth models can be classified into three groups [5, 6, 8]:

(i) Primary growth models describe changes in microbial concentration as a function of time, under defined environmental conditions, allowing the description of microbial curves. In practice, primary growth models assume a sigmoidal growth function, described by three main parameters: maximum growth rate (μ_{max}), lag phase (λ), and maximum population density (Υ_{max}), as shown in Fig. 1. The literature reports a large number of primary models that can be applied to describe microbial growth in food. The most common models are Gompertz and Logistic with their reparameterizations, Baranyi and Roberts, and Buchanan. These models were created and developed based on different theories and

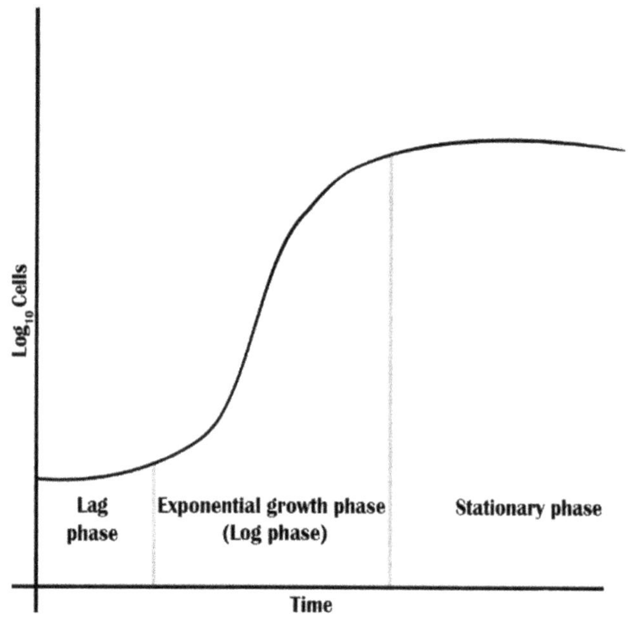

Fig. 1 Typical curve of microbial growth as a function of time

hypotheses, and each one has its advantages and specifications in different applications [9, 10].

(ii) Secondary models describe how parameters obtained from growth models vary as a function of environmental conditions (e.g., temperature, pH, and water activity). Arrhenius, square-root, Gamma concept, artificial neural networks (ANNs), and response surface model are some examples of this type models.

(iii) Tertiary models combine one or more primary and secondary models into user-friendly computational software to provide prediction of microbial growth. Combase, MicroHibro, FSSP, and IPMP 2013 are some software available to facilitate the modeling methodology and are extensively used in food industries.

1.1 Baranyi and Roberts Model

Baranyi and Roberts [11] developed a semi-mechanistic biologically based growth model (Eq. 1 and 2), including an exponential linear growth phase and a lag phase that is calculated by an adjustment function ($A(t)$) (*see* Eq. 2). In this model, the duration of the lag phase will depend on the cell physiology and its adaptation to the new environment, i.e., the lag phase will be longer if the cells are not adapted to the new environment. Baranyi and Roberts model has been widely used in predictive microbiology, due to goodness of fit, and the model parameters are biologically interpretable:

$$\Upsilon = y_0 + \mu_{\max} A(t) - \frac{1}{m} \ln\left(1 + \frac{e^{m\mu_{\max} A(t)} - 1}{e^{m\left(y_{\max} - y_0\right)}}\right) \tag{1}$$

$$A(t) = t + \frac{\ln\left(e^{-\mu_{\max} t} + e^{-h_0} - e^{-v_t - h_0}\right)}{\mu_{\max}} \tag{2}$$

where Υ is the natural logarithm of cell concentration at time t (ln cfu/[g or mL]), y_0 is the natural logarithm of initial bacterial counts (log cfu/[g or mL]), $y_{\max}$ is the natural logarithm of the maximum cell concentration (log cfu/[g or mL]), $\mu_{\max}$ is the maximum specific growth rate (h^{-1}), m is a curvature factor, h_0 is the physiological state of the cells at $t = t_0$, t is the observed time, and v is the rate of increase of the limiting substrate (h^{-1}).

1.2 Gompertz Model

Gompertz model is a double exponential function, which describes an asymmetrical sigmoidal curve. This model is based on the limited space and/or nutrients as well as the production of toxic metabolites, where the growth rate is not constant. In this way, the maximum speed of exponential growth is determined at the inflection point of the curve [9, 11, 12]:

$$\log N(t) = A + Ce^{-e^{(-B(t-M))}} \tag{3}$$

where $N(t)$ is the number of cells (cfu/[g or mL]) at time t *(h)*, A is the asymptotic count as t decreases to 0 (log cfu/[g or mL]), C is the difference in value of the upper and lower asymptote (log cfu/[g or mL]), B is the relative growth rate at M (h^{-1}), and M is the time at which the absolute growth rate is maximum (h).

For biological meaning, the lag time duration and the maximum growth rate can be calculated:

$$\lambda = M - \frac{1}{B} \tag{4}$$

$$\mu = \frac{BC}{e} \tag{5}$$

where e is equal to 2.7182.

1.2.1 Modified Gompertz Model

In some cases, microbial growth data are not described accurately enough by the standard Gompertz model. To assess accuracy, Gompertz model was reparametrized by Zwietering et al. [13]:

$$y = De^{-e^{\left[\frac{\mu e}{A}(\lambda - t) + 1\right]}} \tag{6}$$

where y is the logarithm of the relative population size ($y = \ln[N/N_0]$), N_0 is the microbial concentration (cfu/[mL or g]) at time zero, D is the maximum increase of population attained in the stationary phase, μ is the maximum growth rate (log cfu/[g or mL]/h), and λ is the lag time duration (h).

1.3 Logistic Model

The logistic model is a sigmoidal curve that describes microbial growth as a function of the initial population, time, growth rate, and final population. The logistic model is a sigmoidal function. This model is based on the momentary microbial growth rate, which is proportional to the size population and the material resources in the environment [14, 15]:

$$\log N(t) = A + \frac{C}{1 + e^{-B(t - M)}} \tag{7}$$

where $N(t)$ is the number of cells at time t (log cfu/[g or mL]). The rest of model parameters are as described for Eq. 3.

Such as Gompertz model, for biological meaning, the lag time duration and the maximum growth rate can be calculated:

$$\lambda = M - \frac{2}{B} \tag{8}$$

$$\mu = \frac{BC}{4} \tag{9}$$

1.3.1 Modified Logistic Model

To correct a failure of the logistic model, where the model fit, in many cases, does not start from the point of origin, several modifications have been proposed considering asymmetric growth curves and improving its accuracy [16, 17]. Rosso et al. (1993) proposed a modification in the logistic model with delay ($\lambda > 0$) and without delay ($\lambda = 0$) [18]. This model describes the growth kinetics and estimates generation time and lag time.

$$\text{If } t < \lambda,\ \text{Log}(N_t) = \text{Log}(N_0) \tag{10}$$

$$\text{If } t \geq \lambda,\ \text{Log}\,(N_t) = \text{Log}\left(\frac{N_{\max}}{1 + \left(\frac{N_{\max}}{N_0} - 1\right)e^{-\mu_{\max}(t-\lambda)}}\right) \tag{11}$$

where $N_{\max}$ is the maximum microbial concentration (log cfu/[mL or g]), N_0 is the microbial concentration at time zero (log cfu/[mL or g]), and $\mu_{\max}$ is the maximum specific growth rate (h^{-1}). The rest of model parameters are as described for Eqs. 6 and 7.

1.4 Buchanan Model

Buchanan et al. proposed a three-phase linear model, which is sigmoid and mechanistic. In this model, the growth rate was always at maximum between the end of the lag phase and the start of the stationary phase. The lag phase is divided into two periods: a period of adaptation to the new conditions and the time of energy generation for cell replication [7, 8].

$$\textit{For } t < t_{\text{lag}},\ \log N_t = \log N_0 \quad - \text{lag phase} \tag{12}$$

$$\text{For } t_{\text{lag}} \leq t < t_{\max},\ \log N_t = \log N_0 + \mu(t - t_{\text{lag}}) \quad - \text{exponential growth} \tag{13}$$

$$\text{For } t \geq t_{\max},\ \log N_t = \log N_{\max} \quad - \text{stationary phase} \tag{14}$$

where t_{lag} is the time of the lag phase (h), $t_{\max}$ is the time when the maximum population level is reached (h), and μ is the specific growth rate. The rest of model parameters are as described for Eqs. 1, 6, and 10.

In the following, we will describe the general procedures for estimating microbial growth parameters in artificial culture media or studied food using primary models.

2 Materials

This step is better described in Chapter 2. The main materials used to perform a microbiological analysis of food are described below:

- Stomacher.
- Saline solution (0.85% NaCl).
- Specific culture media (*see* **Note 1**).
- Petri dish.
- Erlenmeyer flask.
- Pipettes (0.1, 1, and 5 mL).
- Laboratory oven (incubator).
- Computer.

3 Methods

Many measurement methods have been developed to enumerate microorganisms in foods and beverages, with the plate count method being the most widely used to describe the microbial growth curve. The growth curve is expressed mainly in terms of microbial number (colony count) as a function of time. For more details on the methods of microbial quantification of foods, *see* Chapter 2.

To develop a primary predictive model, it is necessary (i) to obtain growth data from the microbiological analysis of the study sample [i.e., artificial culture media or studied food, followed by the plate count (*see* Subheading 3.1)], (ii) to graphically represent the growth curve of the microorganism as a function of time resulting from the microbial count data (*see* Subheading 3.2), (iii) to select and fit the primary model to the experimental data to obtain the microbial growth parameters (*see* Subheading 3.3.2), and (iv) to perform statistical analysis (*see* Subheading 3.3.2) (Fig. 2) [19].

3.1 Microbial Analysis

Microbial analysis is the initial step toward a predictive model; it can occur through the use of the natural microbiota of the food itself or through the inoculation of selected pathogens. When the challenge test consists of using the natural microbiota of the food, the steps for microbiological analysis to be followed are sample preparation, incubation, and plate count.

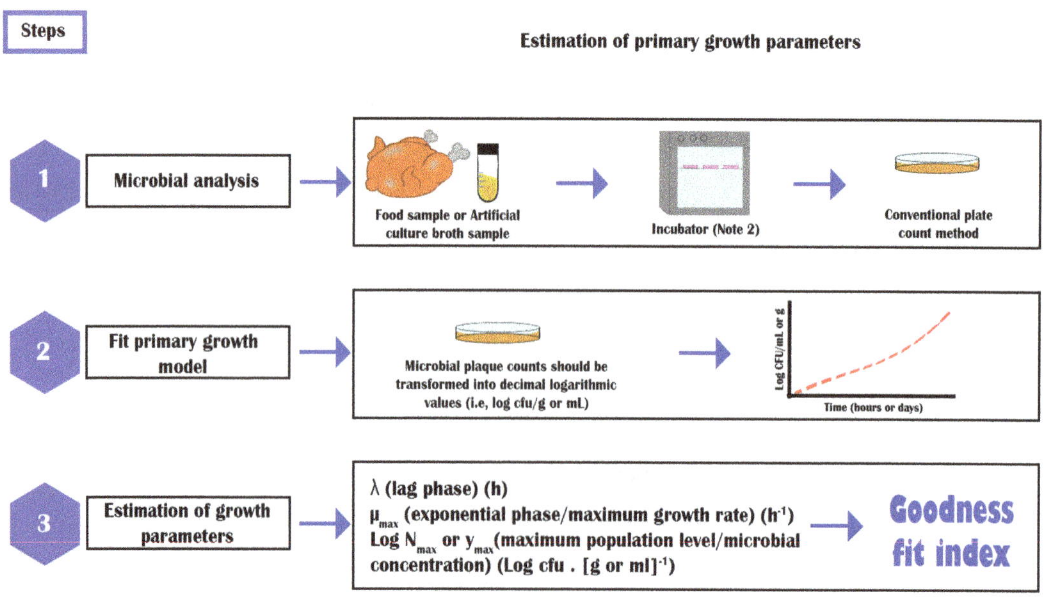

Fig. 2 Stepwise to fit a primary predictive model

For challenge tests of pathogenic bacteria, it will be necessary to inoculate the strain of interest in the food to perform the microbiological analysis. In this case, strains must be clearly identified and characterized, and their origin known [20]. If a surrogate microorganism is used, preliminary work must be done to characterize the strain prior to use (Chapter 2). The microorganisms of interest can be provided by national and international reference laboratories. After obtaining the strain, the inoculum must be prepared according to the supplier's instructions. Inoculum preparation will depend on the species, strains, cell type (vegetative or spores), objectives of the study, etc. (*see* **Note 3**).

3.2 Growth Curve Experiments

For the development of the growth curve, plaque counts (colony forming units; CFU) should be converted to decimal logarithmic values ($\log_{10}$). After correcting the data, the microbial growth curves should be plotted taking into account the microbial concentration over time. The method selected for building growth curves will depend on the resources available for testing [21]. For more information concerning methods of quantification microbial response, *see* Chapter 2.

3.3 Primary Growth Modeling

This step consists of analyzing the growth curves using mathematical models. As previously mentioned, computational tools can fit and estimate microbial growth parameters from the curves obtained from the plate count. Currently, there are many computational tools for fitting primary models. These devices can be separated into two categories, online and offline. For more information about software, *see* Chapter 1. From this point, we provide a stepwise for fit primary models in DMFit online and offline versions.

3.3.1 DMFit

DMFit seeks to fit microbial curves where a lag phase is preceded and followed by a linear phase and a stationary phase. It can be used to visualize a graphical representation of microbiological growth data and fit a growth model to the data. In DMFit, microbial curves can be fitted to two different types of models: Baranyi and Roberts model and the trilinear, biphasic, and linear models. This free tool is available in two versions: online (DMFit of ComBase) and offline (DMFit add-in for Excel).

3.3.1.1 DMFit: Online Version

To use DMFIT online, it is necessary to register for free accessing the ComBase website (https://browser.combase.cc/). As shown in Fig. 3, the first step consists to click on "DMFit" and then on "online DMFit" (*a*). Then, two columns will appear, corresponding to time (days or hours) and microbial count ($\log_{10}$) (*b*). Finally, press "Fit" (*c*) and select the model of interest (*d*).

Once the selected model has been fitted, DMFit online allows the user to obtain estimates of growth parameters such as

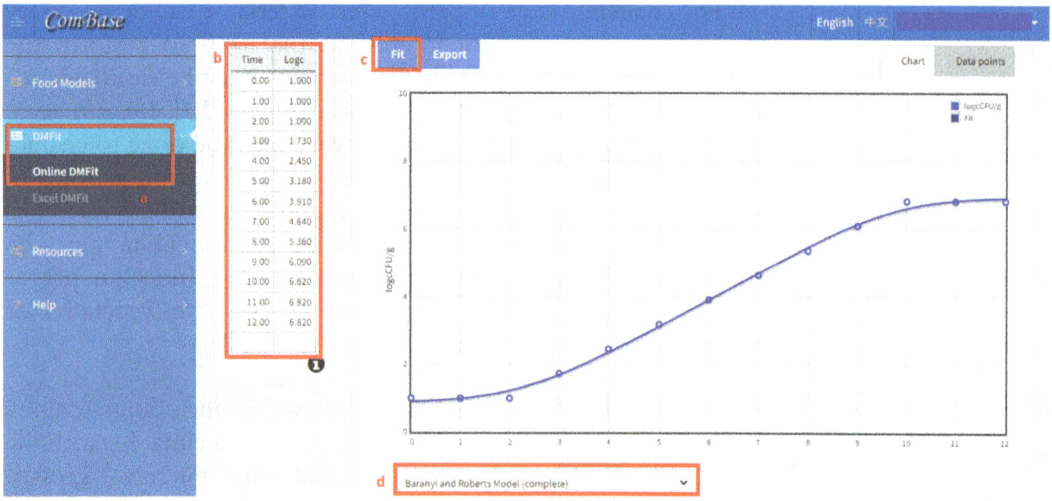

Fig. 3 Stepwise to fit microbial growth models to experimental data using Online DMFit

R-square:	0.997
SE of Fit:	0.126
Initial value	0.902 ± 0.0906
Lag/shoulder	2.283 ± 0.22
Maximum Rate	0.807 ± 0.0293
Final Value	6.938 ± 0.103

Fig. 4 Estimated growth parameters and statistical indexes obtained from the fit of a primary model

maximum growth rate (h^{-1}), lag phase (lag/shoulder) (hours or days), initial and final cell count, and estimation of standard errors of these parameters. In addition, the tools provide statistical indexes, based on the coefficient of determination (R^2) and the standard error of fit (SE of fit) (Fig. 4). Based on the selected model, estimated values for growth parameters and statistical analysis may vary (*see* **Note 4**).

3.3.1.2 Excel DMFit: Offline Version

The offline version of DMFit is an Excel add-in that fits a microbial growth model to count versus time. This version also estimates growth parameters such as lag phase, growth rate, and maximum microbial concentration.

To use DMFit as an Excel add-in, access the ComBase website (https://browser.combase.cc/) and click on "DMFit," followed by "Excel DMFit," as shown in Fig. 5a. Then, to download the add-in, click on "DMFit for Excel" (Fig. 5b).

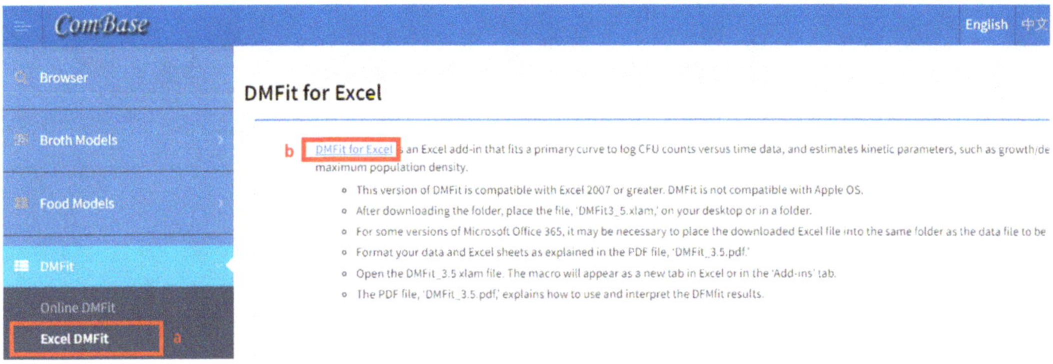

Fig. 5 Stepwise to download DMFit as an Excel add-in

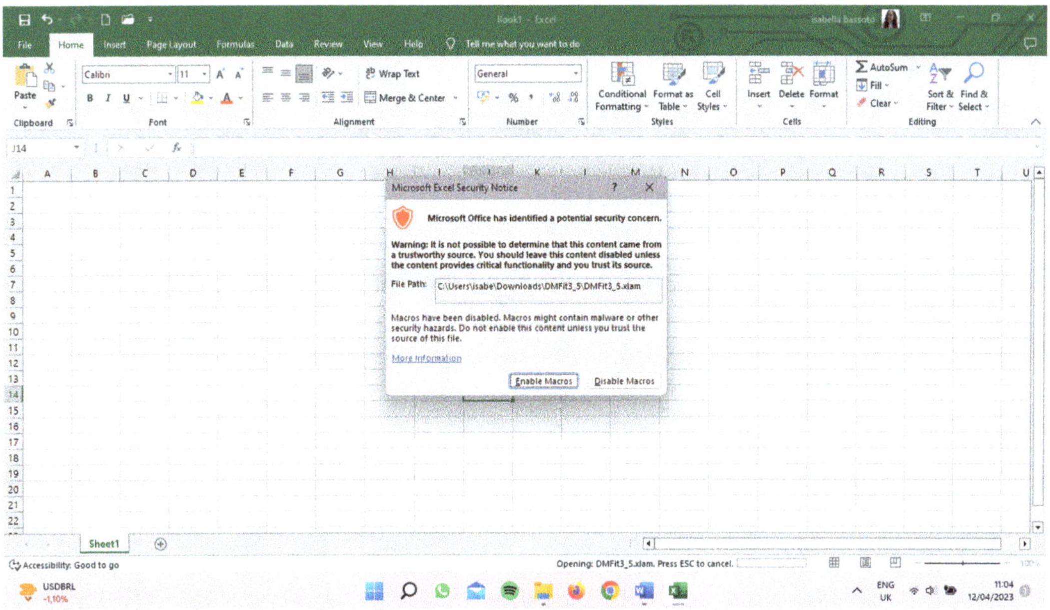

Fig. 6 Enable Macros for the correct use of DMFit offline version

After downloading DMFit, extract the zip folder and click on Excel file "DMFit3_5." To start using the tool, the user must click on "Enable Macros," as shown in Fig. 6.

Next, insert the time and microbial concentration data (Fig. 7). The microbial count data must be in $\log10_{10}$, and when the log values are 0, the value 0.01 must be adopted.

Once the data has been entered, select only the values of the columns of time and microbial concentration, and in the toolbar, click on "DMFit" (Fig. 8a). Then click on "Fit curve defined by selection" (Fig. 8b).

Once the model is fitted to the experimental data, the DMFit will automatically open a new spreadsheet with the curve fitting and estimated growth parameters. These parameters in DMFit as an

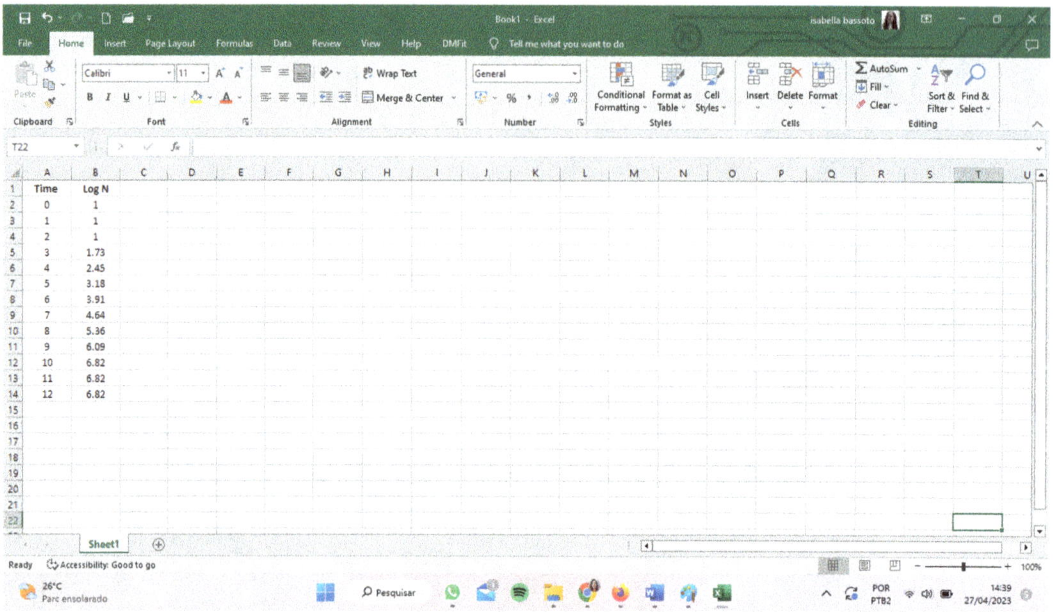

Fig. 7 Input of time and microbial concentration data

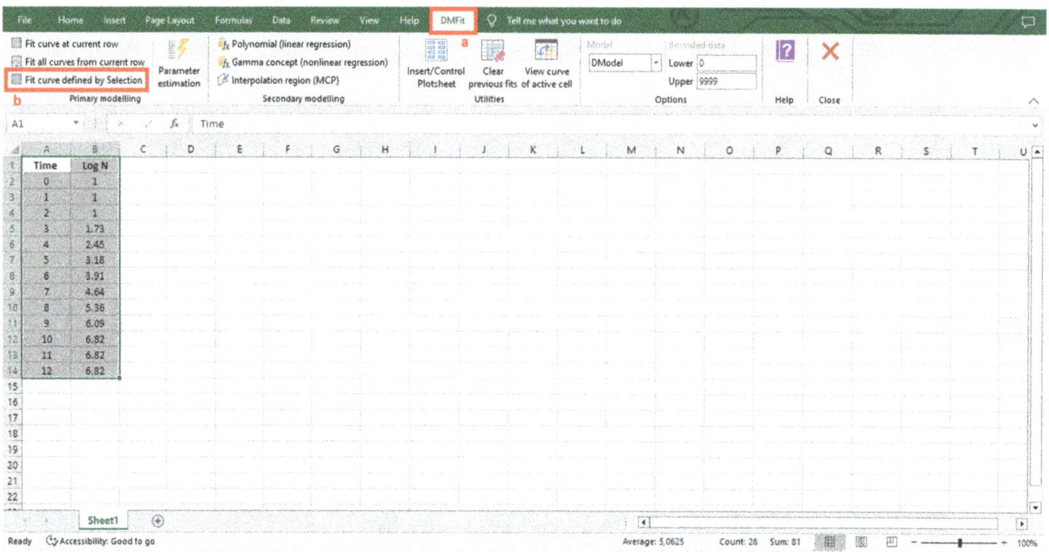

Fig. 8 Construction of microbial growth curve as a function of time

Excel add-in are defined as follows: "rate" (growth rate), lag (lag phase), *y*0 (initial population), and maximum population (yEnd). In addition, DMFit provides error code of the model (errCode), coefficient of determination (R^2), and the curvature parameters, described by *mCurve* and *nCurve* (Fig. 9) (*see* **Notes 5** and **6**).

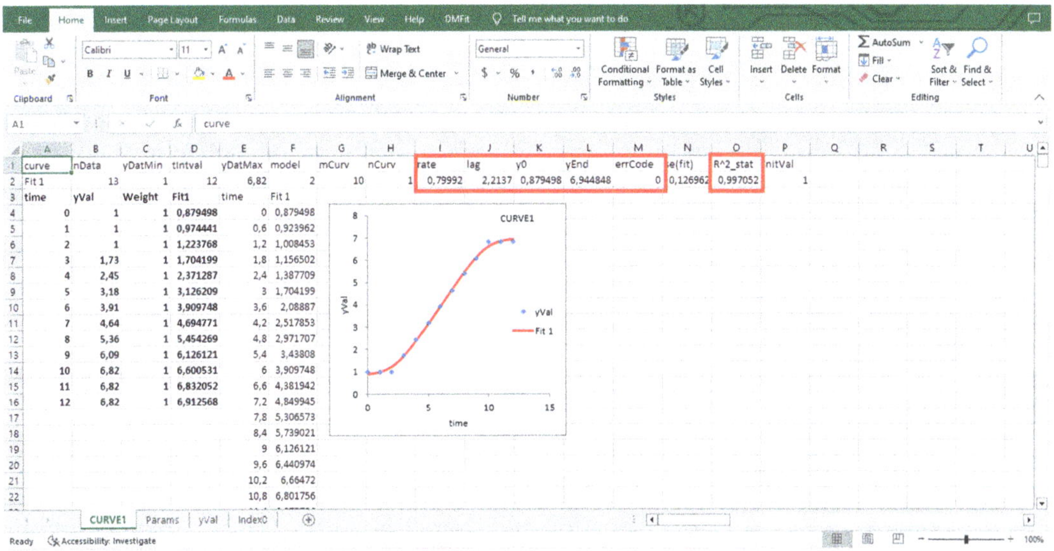

Fig. 9 Fitting the model to experimental data and estimating growth parameters

3.3.2 Goodness of Fit Index

The performance of the primary growth models can be evaluated using statistical indexes such adjusted coefficient of determination ($R^2{}_{\text{adj}}$), residual sum of squares (RSS), and root mean square error (RMSE), described by Eqs. 15, 16, 17, and 18, respectively [22]:

$$\text{RSS} = \sum_{i=1}^{n} (y_i - \widehat{y})^2 \tag{15}$$

$$\text{RMSE} = \sqrt{\frac{\text{RSS}}{n-k}} \tag{16}$$

$$R^2 = \frac{\sum_{i=1}^{n} \left(\widehat{y}_i - \frac{1}{n} \sum_{i=1}^{n} y_i \right)^2}{\sum_{i=1}^{n} \left(y_i - \frac{1}{n} \sum_{i=1}^{n} y_i \right)^2} \tag{17}$$

$$R^2_{\text{adj}} = 1 - \left(1 - R^2\right) \frac{n-1}{n-k-1} \tag{18}$$

where y_i is the observed value for sample i, $\widehat{y}_i$ is the value of the prediction for sample i, k is the number of parameters, and n is the number of sample data.

4 Limitations

Mathematical models have some limitations that must be considered. The models fit the observed data. Therefore, they cannot be extrapolated outside the ranges in which they were derived.

Predictions outside the experimental ranges are generally inaccurate and, in some cases, insecure [23].

Models are usually conducted in laboratory media and generally predict faster growth rates than observed. This makes them fail-safe, and although they are validated in food, they may not have widespread application in the food industry [23, 24]. Another limitation is that models derived under static conditions may not be applicable to fluctuating conditions [23, 24].

5 Notes

1. Medium selection will depend on the microorganism studied. The "Compendium of Methods for the Microbiological Examination of Foods: APHA" can be consulted to assist in this step.
2. The incubation time varies according to the microorganism studied and its methodology. The "Compendium of Methods for the Microbiological Examination of Foods: APHA" can be consulted to assist in this step.
3. To understand about the inoculation and sample preparation step, Chapter 2 can be consulted.
4. For Baranyi and Roberts model (complete), the growth parameters estimated are λ, μ_{max}, y_{max}, and y_0. The Baranyi and Roberts model (no lag) estimates μ_{max}, y_{max}, and y_0. Baranyi and Roberts model (no asymptote) estimates λ, μ_{max}, and y_0. The trilinear model growth parameters are μmax, λ, y_0, and time when the maximum population density is reached (t_{max}). Biphasic model (no lag) estimates t_{max}, y_0, and μ_{max}. For biphasic model (no asymptote), the growth parameters are $y0$, λ, and μ_{max}. The growth parameters for linear model are $y0$ and μ_{max}.
5. For each error code value, there is an explanation, where *(a)* 0 – model within standards; *(b)* 1 – discontinuous problem: discontinuity problem in the model (straight-line model), or gear = 0; *(c)* 2 – some parameters have very high standard error (rel. error > 200%); *(d)* 3 – values do not fit any model (ill conditioned problem); *(e)* 4 – modeling led to very high or very low values (running into singularities during the calculations); *(f)* 5 – error in input data in the index sheet or error in the raw data of the curves; and *(g)* 6 – the curve to be fitted is not found in the actual directory.
6. For curvatures after the lag phase, and after the exponential phase, the program defaults to *mCurv* = 10 and *nCurv* = 1. However, you must adjust the *mCurv* and *nCurv* values and compare them with the program's default, by setting R^2. The user has the possibility to improve the curve fit from *mCurve nCurve* trials. For each new attempt, a new curve will appear on the worksheet.

References

1. Food and Drug Administration (FDA). What you need to know about foodborne illnesses. https://www.fda.gov/food/resourcesforyou/consumers/ucm103263.htm
2. Cho IH, Ku S (2017) Current technical approaches for the early detection of food-borne pathogens: Challenges and opportunities. Intern J Molec Sci 18(10):2078. MDPI AG. https://doi.org/10.3390/ijms18102078
3. Moraes JO, Cruz EA, Pinheiro Í, Oliveira TCM, Alvarenga V, Sant'Ana AS, Magnani M (2019) An ordinal logistic regression approach to predict the variability on biofilm formation stages by five Salmonella enterica strains on polypropylene and glass surfaces as affected by pH, temperature and NaCl. Food Microbiol 83:95–103. https://doi.org/10.1016/j.fm.2019.04.012
4. Souza PBA, Poltronieri KF, Alvarenga VO, Granato D, Rodriguez ADD, Sant'Ana AS, Peña WEL (2017) Modeling of Byssochamys nivea and Neosartorya fischeri inactivation in papaya and pineapple juices as a function of temperature and soluble solids content. LWT Food Sci Technol 82:90–95. https://doi.org/10.1016/j.lwt.2017.04.021
5. Baranyi J, Pin C (2001) A parallel study on bacterial growth and inactivation. J Theor Biol 210(3):327–336. https://doi.org/10.1006/jtbi.2001.2312
6. Ross T, Mcmeekin TA (2003) Modeling Microbial Growth Within Food Safety Risk Assessments. Risk Anal 23(1):179
7. Whiting RC (1995) Microbial modeling in foods. Crit Rev Food Sci Nutr 35(6):464–494
8. Perez-Rodriguez F, Valero A (2013) Predictive microbiology in foods. SpringerBriefs Food, Health Nutr. https://doi.org/10.1007/978-1-4614-5520-2
9. McKellar RC, Lu X (2003) Modelling microbial responses in food. CRC Series in Contemporary Food Science. CRC, London. ISBN 0-8493-1237-X
10. Mahdinia E, Liu S, Demirci A, Puri VM (2020) Microbial Growth Models. In: Demirci A, Feng H, Krishnamurthy K (eds) Food Safety Engineering. Food Engineering Series. Springer, Cham. https://doi.org/10.1007/978-3-030-42660-6_14
11. Baranyi J, Roberts TA (1994) Review Paper A dynamic approach to predicting bacterial growth in food. Intern J Food Microbiol 23:277–294
12. Stavropoulou, E., & Bezirtzoglou, E. (2019). Predictive modeling of microbial behavior in food. Foods, 8(12). MDPI Multidisciplinary Digital Publishing Institute. https://doi.org/10.3390/foods8120654
13. Zwietering MH, Jongenburger I, Rombouts FM, van't Riet K (1990) Modeling of the bacterial growth curve. Appl Environ Microbiol 56:1875
14. Peleg M, Corradini MG, Normand MD (2007) The logistic (Verhulst) model for sigmoid microbial growth curves revisited. Food Res Int 40(7):808–818. https://doi.org/10.1016/j.foodres.2007.01.012
15. Wachenheim DE, Patterson JA, Ladisch MR (2003) Analysis of the logistic function model: derivation and applications specific to batch cultured microorganisms. Bioresour Technol 86(2):157–164. https://doi.org/10.1016/S0960-8524(02)00149-9
16. Augustin J-C, Carlier V (2000) Mathematical modelling of the growth rate and lag time for Listeria monocytogenes. Int J Food Microbiol 56:29–51
17. Fujikawa H, Kai A, Morozumi S (2003) A new logistic model for bacterial growth. J Food Hygien Soc Japan (Shokuhin Eiseigaku Zasshi) 44(3):155–160. https://doi.org/10.3358/shokueishi.44.155
18. Rosso L, Lobry JR, Flandrois JP (1993) An Unexpected correlation between cardinal temperatures of microbial growth highlighted by a new model. J Theor Biol 162(4):447–463. https://doi.org/10.1006/jtbi.1993.1099
19. Mytilinaios I, Salih M, Schofield HK, Lambert RJW (2012) Growth curve prediction from optical density data. Int J Food Microbiol 154(3):169–176. https://doi.org/10.1016/j.ijfoodmicro.2011.12.035
20. Sinclair RG, Rose JB, Hashsham SA, Gerba CP, Haase CN (2012) Criteria for selection of surrogates used to study the sate and control of pathogens in the environment. Appl Environ Microbiol 78(6):1969–1977. https://doi.org/10.1128/AEM.06582-11
21. Zhang X, Jiang X, Hao Z, Qu K (2019) Advances in online methods for monitoring microbial growth. Biosens Bioelectron 126:433–447. Elsevier Ltd. https://doi.org/10.1016/j.bios.2018.10.035
22. Liu Y, Wang X, Liu B, Dong Q (2020) Microrisk Lab: an online freeware for predictive microbiology 1. https://doi.org/10.1101/2020.07.23.218909
23. Thakur B, Amarnath CA, Mangoli SH, Sawant SN (2015) Polyaniline nanoparticle based colorimetric sensor for monitoring bacterial

growth. Sens Actuators B Chem 207(Part A):262–268. https://doi.org/10.1016/j.snb.2014.10.045

24. Weidemaier K, Carruthers E, Curry A, Kuroda M, Fallows E, Thomas J, Sherman D, Muldoon M (2015) Real-time pathogen monitoring during enrichment: A novel nanotechnology-based approach to food safety testing. Int J Food Microbiol 198:19–27. https://doi.org/10.1016/j.ijfoodmicro.2014.12.018

Chapter 5

A Protocol for Predictive Modeling of Microbial Inactivation Based on Experimental Data

Leonidas Georgalis, Pablo S. Fernandez, and Alberto Garre

Abstract

Temperature treatments are one of the most commonly used methods for microbial inactivation in food industries. Microbial inactivation models can support optimization of thermal treatments, due to their ability to predict the response of microbial populations under fixed temperature conditions (isothermal conditions) or variable (dynamic conditions). These models can also consider intrinsic and extrinsic factors involved in the process. This chapter aims to describe a protocol for developing models for thermal inactivation using conceptual data from the literature and empirical experiments. The online and open-source interface bioinactivation (available at https://foodlab-upct.shinyapps.io/bioinactivation4/) was used for the computations. The protocol includes conceptual and empirical phases and defines aspects from elaborating a reasonable hypothesis and selecting commonly used equations to the model fitting and validation.

Key words Predictive microbiology, Parameter estimation, Model fitting, Simulation, Mathematical modeling, Model validation

1 Introduction

Inactivation treatments are essential for food safety. Most bacterial pathogens are ubiquitous in the environment (e.g., *Bacillus cereus* or *Listeria monocytogenes*) or in the intestine of animals (e.g., *Campylobacter* spp. or *Escherichia coli*), making them present in many food ingredients [1]. Moreover, many of them can survive within processing environments (e.g., as biofilms), so they may be able to enter the food chain during production [2]. For that reason, it is paramount for the safety of most products the introduction of an inactivation step that reduces the microbial concentration [3]. Although many novel inactivation technologies have been suggested during the last decades [4], the application of high temperatures remains one of the technologies most commonly used by food industries. Temperatures in the range of 72 °C have demonstrated their ability to inactivate vegetative microbial cells,

Verônica Ortiz Alvarenga (ed.), *Basic Protocols in Predictive Food Microbiology*, Methods and Protocols in Food Science,
https://doi.org/10.1007/978-1-0716-3413-4_5,

whereas temperatures usually above 120 °C are used to inactivate bacterial spores [5]. On the other hand, high temperatures can also have deleterious effects on the food product, as they can accelerate or cause the denaturation or decomposition of health-promoting compounds present in the food (e.g., vitamins, antioxidants, or probiotic bacteria) [6, 7]. Furthermore, the application of the thermal treatment may require large amounts of energy, increasing the carbon footprint of food production. Therefore, the design of effective thermal treatments (i.e., the treatment temperature and its duration) must balance food safety, food quality, and environmental impact.

Mathematical models have proved to be effective tools to assist process design [3]. Predictive models can estimate the number of microbial cells surviving a treatment of duration t at temperature T. Hence, they can be useful when designing thermal treatments able to cause the desired microbial inactivation with a minimal impact on the quality attributes of the product and its environmental impact [8, 9]. Although the first scientific studies dedicated to modeling microbial inactivation date back to the 1920s [10–12], these tools were only sparingly used in industry at the time. It was not until the last few decades that they were broadly adopted, probably due to the availability of efficient computation methods. Nowadays, predictive models can be considered standard tools to support process design, shelf life estimation, and risk assessment [13, 14].

Although general conventions in food microbiology about the model equations used to describe microbial inactivation have been stablished, their model parameters are mostly unknown. Inactivation rates depend on external process parameters (e.g., temperature), extrinsic parameters (e.g., pH or aw of the product, presence of antimicrobials, etc.), and intrinsic parameters (e.g., bacterial strain or preculture conditions) [15]. These relationships have not yet been described mechanistically, so they must be estimated based on experimental data [16]. In this sense, this chapter proposes a protocol to build predictive models for microbial inactivation. The protocol (described in Subheading 2) covers both conceptual and practical (experimental and computational) aspects. The reason for this is that inactivation models from predictive microbiology are always empirical, so they can never be better than the data used to build them. In other words, the statistical method cannot extract information not contained in the data, regardless of the complexity of the analysis method. The computational parts of the protocol are based on the web version of *bioinactivation* (*see* Chapter 1; available online at https://foodlab-upct.shinyapps.io/bioinactivation4/) [17, 18]. It is followed in Subheading 3 by a description of the equipment and materials that are commonly needed for microbial inactivation experiments. Finally, Subheading 4 emphasizes some common issues and important points to consider when developing microbial inactivation models. Although this chapter is focused on thermal microbial inactivation, most of the principles presented here are also applicable to the development of predictive models for non-thermal technologies. Furthermore, the concepts discussed here can also be applied to develop inactivation models where the response variable is not the microbial concentration (e.g., the loss of quality attributes).

2 Protocol for Building Predictive Models for Heat Inactivation

The protocol is summarized in Fig. 1. It consists of seven steps, divided in two phases (a conceptual phase and a practical phase) that precede the application of the model to actual case studies.

The first one is the *conceptual phase* (*see* **Note 1**), which includes all the activities required to frame and plan the study.

- The first step is the *definition of the scope* of the study. Microbial inactivation is a very broad subject that cannot be fully covered in a single study. For that reason, it is important to clearly define the scientific question before taking the next steps of the study.
- The second step is to perform a *literature review* that should support the rest of the model building process. This includes a review of previously published studies (general information about the microorganism(s) of interest, relevant factors for the model, etc.), as well as the candidate model equations.
- The third step is *model definition*. Based on the scope of the study and the information available in the literature, one should define a list of candidate model equations that are likely to describe the experimental data.

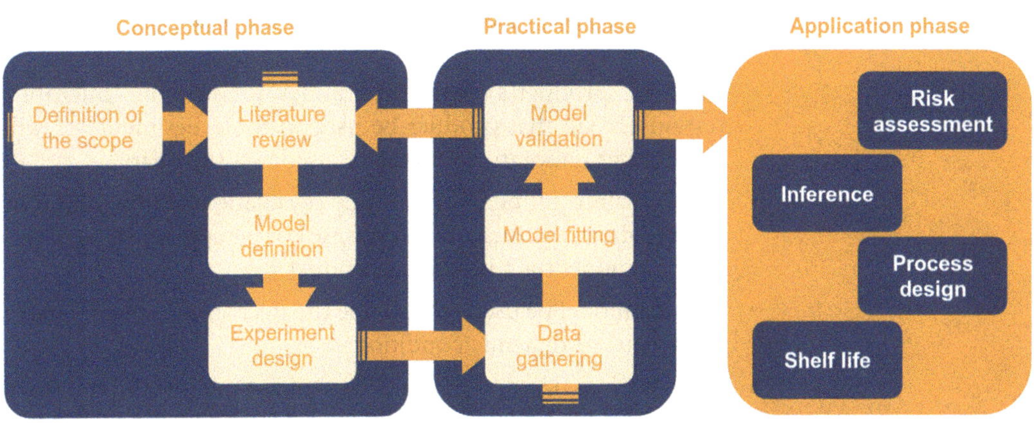

Fig. 1 Diagram illustrating the proposed protocol to build microbial inactivation models

- The fourth step is the elaboration of an *experimental design* that should provide enough information to build the model based on the information gathered on the previous step.

The second phase of the protocol is the *practical phase*, which covers the experimental and data analysis parts of the study.

- The first step within the practical phase is *data gathering*. This step involves carrying out the inactivation experiments that were designed in the previous steps.
- Once the data have been gathered, the most likely values of the model parameters are estimated by *model fitting* using the most suitable algorithm.
- Before the model is ready to be applied, it must go through *model validation*. This includes the evaluation of the model from a statistical point of view, as well as the comparison against independent data (either from literature or generated within the frame of the study). As illustrated in Fig. 1, the results of this step will decide whether the model is ready to enter the application phase or one needs to return to the conceptual phase.

The following subsections detail each one of these steps.

2.1 Definition of the Scope

Microbial inactivation is a very broad subject that depends on a large variety of factors, making the definition of a general mathematical model that is applicable for all cases impossible. For that reason, it is important to define in detail the scope of the model before any other step is taken. This includes but is not limited to the following:

- What is the target microorganism(s)? Will the model be specific to a particular bacterial strain, or will it account for the variability in the microbial response of a taxonomic unit (species, family, etc.)?
- In what product(s) will the model predict the microbial response? Will that product be used in the experiments or will some other (laboratory) media be used as a surrogate?
- Is this mainly a fundamental study (evaluation of how different aspects affect the microbial response) or an applied one (developing a model for a particular product)?
- Will the model be used to describe the microbial response under isothermal or dynamic conditions?
- What intrinsic factors (pH, presence of antimicrobials, etc.) may be relevant for the case studied?
- What process factors (heating rate, direct or indirect heating, aspects linked to technology, etc.) may have a relevant role in the microbial response?

- What other factors (culture conditions, recovery conditions, etc.) may be relevant for the interpretation of the results?

These and other relevant questions should be carefully considered before the next steps of the study are taken.

2.2 Literature Review

The kinetics of microbial inactivation have been a focus of scientific studies for decades, and most models used in predictive microbiology have been available for over two decades. For that reason, it is advisable to begin the study with a detailed review of the data already available in the scientific literature, as this information will be valuable in subsequent steps of the protocol, from experimental design to model validation. The literature review should cover, at least, the following aspects:

- Have previous studies analyzed microbial inactivation under similar conditions (species, media, etc.)?
- Is it possible to have an indication of the expected microbial response (D-value, time to six log-reductions, etc.)?
- What is the expected effect of intrinsic or other process factors on the microbial response (e.g., effect of pH, type of acids, etc. on the D-value)?
- What method(s) was used to establish the microbial inactivation in previous studies? Are there possible limitations in those studies (e.g., come-up times not accounted for)?
- What mathematical models are available for this process? Are nonlinearities (e.g., shoulders or tails) expected?
- What are the most appropriate preculture and recovery conditions? Would the use of selective media provide valuable information?

This type of information can be gathered from various sources. General information on microbial inactivation can be retrieved from common online repositories (Scopus, Google Scholar, etc.) or scientific books on food microbiology. Moreover, there are several online databases for predictive microbiology (ComBase, https://www.combase.cc; Sym'previus, https://symprevius.eu; D-database, https://foodmicrowur.shinyapps.io/Ddatabase/; MicroHibro, https://microhibro.com/ etc.) that can provide valuable information for this kind of studies [19].

2.3 Model Definition

The third step in the protocol is the definition of the (candidate) model equation(s). It is crucial in this step that the models are defined according to the goal(s) of the study. For instance, if the aim is to analyze whether an antimicrobial affects the inactivation rate, models with and without this effect should be considered. Moreover, the models should, when possible, account for any relevant information retrieved during the literature review.

Although it may be more common to introduce this step later in the protocol, the definition of a list of candidate models can be useful to design experiments that will be better able to discriminate between these model structures [20]. Consequently, we encourage researchers to consider the candidate model equations before the experimental design. We also strongly recommend that (most of) the proposed models are based on model equations and parameterizations that are commonly used in predictive microbiology (listed below). There are two reasons for this. The first one is that these models are easily understood by other food scientists, making communication easier. The second one is that comparison to similar studies from the literature is a crucial step of model development (reasoned in Subheading 2.7); this comparison is much more simple when both studies use the same model equations. Another recommendation for model definition is to always include candidate models based on simple hypotheses (log-linear relationships, models only considering treatment time and temperature). Even when they are expected to fail at describing the experimental data, these deviations are a good justification for the use of more complex models [21]. Furthermore, reporting the limitations of the simplified model can help when comparing against the results of other scientific studies.

The following two sections describe the model equations most commonly used in predictive microbiology to describe thermal inactivation. We focus on the effect of temperature on the microbial response, as the influence of other factors (e.g., pH or concentration of antimicrobials) involves many factors and would be lengthy to describe. Because there are fundamental differences between isothermal and dynamic conditions for model definition and parameter estimation, they are explained in separate sections here. Subheading 2.3.3 will comment on how these models can be extended to include additional environmental factors.

2.3.1 Inactivation Models for Isothermal Conditions

Predictive microbiology studies focused on isothermal conditions usually divide model definition in two steps [13, 22]. First, an equation to describe the relationship between the microbial count and the elapsed time is proposed (*primary model*). Primary models usually have model parameters that depend on the environmental conditions, a typical example being the relationship between the inactivation rate and the treatment temperature (inactivation is faster at higher temperatures). Then, *secondary models* describe how the model parameters of the primary model are affected by changes in the environmental conditions.

Probably the most simple primary model to describe microbial inactivation is a log-linear relationship between the microbial count (N) and the treatment time (t), as shown in Eq. (1) where "log" refers to the decimal logarithm (this convention is kept within the whole chapter).

$$\log N(t) = \log N_0 - \frac{t}{D(T)} \tag{1}$$

This model is described by two parameters: the microbial count at the beginning of the treatment (N_0) and the D-value ($D(T)$). This parameter defines the velocity of the inactivation, being equal to the treatment time required to inactivate 90% of the microbial cells in the population at temperature T. The effect of temperature changes on the inactivation rate is described using a secondary model that considers a log-linear relationship between the D-value and the treatment temperature (Eq. 2). This equation introduces the so-called z-value (z) that defines the temperature increase required to cause a tenfold reduction of the D-value. This model also introduces a reference temperature (T_{ref}), without biological meaning but with a positive impact on parameter identifiability and interpretability [23, 24]. Then, the secondary model is described by two parameters: the z-value and the D-value at the reference temperature (D_{ref}). Although there is discussion whether these model equations can be attributed to the studies by Bigelow in the 1920s, many food microbiologists refer to them as the "Bigelow model" as a recognition of their pioneering studies.

$$\log D(T) = \log D_{\text{ref}} - \frac{T - T_{ref}}{z} \tag{2}$$

Several studies have shown that the microbial response to isothermal treatments usually deviates from the simple hypotheses of the Bigelow model. Consequently, two main approaches have been developed to describe more complex responses: the "vitalistic" and the "mechanistic" [25]. The vitalistic approach considers that the stress resistance of the microbial cells varies within the population according to some distribution (usually Weibull) [26]. This hypothesis has been applied in three equivalent primary models with different parameterizations. The one by Peleg and Cole [26] uses the inactivation rate ($b(T)$), as shown in Eq. (3), whereas the one by Mafart et al. [27] uses the "delta-value" ($\delta(T)$) which equals the treatment time required to cause the first log-reduction in the microbial population (Eq. 4). The nonlinearity of the survivor curve is described by the shape factor of the Weibull distribution (n in the parameterization by Peleg [26], p in the parameterization by Mafart [27]). Values of this parameter greater than one introduce a downward curvature of the survivor curve, whereas values lower than one result in curves with an upward curvature. In the particular case where this parameter equals one, the survivor curves are linear, and the model predictions are equivalent to those of the Bigelow model.

$$\log N(t) = \log N_0 - b(T) \cdot t^n \tag{3}$$

$$\log N(t) = \log N_0 - \left(\frac{t}{\delta(T)}\right)^p \tag{4}$$

Note that parameters b and δ are related by the identity $b = 1/\delta^p$, so both primary models are equivalent. However, these models differ in their secondary models. Peleg and Cole [26] propose the relationship between $b(T)$ and temperature (T) described in Eq. (5). It assumes that, for temperatures lower than a critical temperature (T_c), the inactivation rate is zero (i.e., there is no inactivation). Then, for higher temperatures, there is a linear relationship between b and T with a superlinear transition for temperatures close to T_c.

$$b(T) - \ln\{1 + [(k_b(T - T_c)]\} \tag{5}$$

The Mafart model assumes instead a log-linear relationship between $\delta(T)$ and temperature, similar to the one hypothesized in the Bigelow model (Eq. 6). Hence, this secondary model is also described by the z-value (z) and the value of $\delta(T)$ at the reference temperature (δ_{ref}).

$$\log \delta(T) = \log \delta_{\text{ref}} - \frac{T - T_{\text{ref}}}{z} \tag{6}$$

Regarding the shape parameter of the Weibull distribution (n or p, depending on the parameterization), it is commonly assumed in both modeling approaches that it does not depend on temperature or any other environmental factors [28].

Recently, a variation in the parameterization of the Mafart model was suggested [29]. Instead of using the "delta-value" (the treatment time for one log-reduction), it uses the parameter D_Δ, as shown in Eq. (7). In this model, the product of Δ and D_Δ equals the treatment time required to cause Δ log-reduction of the microbial count ($t_\Delta = \Delta \cdot D_\Delta$). This model has the advantage that parameter D_Δ is usually more relevant than the delta-value, as Δ can be set to the target log-reduction for the treatment (e.g., five or six log-reductions). Furthermore, this reparameterization can reduce the collinearity between parameters p and δ, reducing parameter uncertainty. This model uses a similar secondary model for D_Δ as the one used in the Mafart model (Eq. 8).

$$\log N(T) = \log N_0 - \Delta\left(\frac{t}{\Delta \cdot D_\Delta(T)}\right)^p \tag{7}$$

$$\log D_\Delta(T) = \log D_{\Delta\text{ref}} - \frac{T - T_{\text{ref}}}{z} \tag{8}$$

Instead of assuming that stress resistance of individual cells follows a probability distribution, models with a mechanistic basis assume that deviations in the inactivation rate are the result of population level phenomena. The most common mechanistic model for microbial inactivation is the one proposed by Geeraerd et al. [30], developed based on hypotheses similar to the ones used in the Baranyi growth model [31]. It can be seen as an extension of a first-order reaction, where the inactivation rate (k) equals the product of three terms ($k = a{\cdot}k'{\cdot}\beta$), as shown in Eq. (9).

$$\frac{dN}{dt} = \alpha(t).k'.\beta(t).N(t) \tag{9}$$

The terms $a(t)$ and $\beta(t)$ are correction factors taking values between 0 and 1 that describe, respectively, the shoulder (delay at the beginning of the treatment) and tail (stabilization at the end of the treatment) of the survivor curve. Therefore, this model is able to describe sigmoidal curves, with two inflection points.

The shoulder is modeled as shown in Eqs. (10 and 11). It is based on the assumption of an ideal substance ($C(t)$) that must be inactivated before microbial inactivation takes place. This model assumes that $C(t)$ follows first-order kinetics with rate k'.

$$\alpha(t) = \frac{1}{1 + C(t)} \tag{10}$$

$$\frac{dC}{dt} = -k' \cdot C(T) \tag{11}$$

Hence, the shoulder length (SL) is defined by the initial value of $C(t)$ ($C(0)$) according to Eq. (12):

$$\mathrm{SL} = \frac{\ln C(0) + 1}{k'} \tag{12}$$

The tail is described in this model using a Verhulst logistic term (Eq. 13), where N_{res} is the tail height.

$$\beta(t) = 1 - \frac{N_{\mathrm{res}}}{N(t)} \tag{13}$$

Under isothermal conditions, the Geeraerd model can be written as shown in Eq. (14):

$$\log N(t) = \log N_{\mathrm{res}} + \log\left(\left(\left(10^{\log N_0 - \log N_{\mathrm{res}}} - 1\right).e^{k'\mathrm{SL}}\right)/\left(e^{k'.t} + e^{k'\mathrm{SL}} - 1\right) + 1\right) \tag{14}$$

In order to ease comparison with other modeling approaches, it is common to analyze the Geeraerd model based on the D-value during the log-linear phase instead of k. Considering that both parameters are related by the identity $D(T) = \ln(10)/k'$, a secondary model for this parameter can be proposed based on the same hypotheses as those in the Bigelow model (Eq. 10). Regarding the two other model parameters (SL and N_{res}), there is no clear consensus on how to describe their relationship with temperature. Therefore, this relationship must be analyzed in detail for each case study.

Table 1
Common models for describing microbial inactivation under isothermal conditions

Model name	Primary model	Secondary model
Bigelow	$\text{Log } N(t) = \log N_0 - t/D(T)$	$\text{Log } D(T) = \log D_{ref} - (T - T_{ref})/z$
Peleg	$\text{Log } N(t) = \log N_0 - b(T) \cdot t^n$	$b(T) = \ln(1 + \exp.(k_b(T - T_c)))$
Mafart	$\text{Log } N(t) = \log N_0 - (t/\delta(T))^p$	$\text{Log } \delta(T) = \log \delta_{ref} - (T - T_{ref})/z$
Metselaar	$\text{Log } N(t) = \log N_0 - \Delta(t/D_\Delta(T))^p$	$\text{Log } D_\Delta(T) = \log D_{\Delta,ref} - (T - T_{ref})/z$
Geeraerd	$\text{Log } N(t) = \log N_{res} + \log(((10^{\log N0 - \log Nres} - 1) \cdot e^{k' \cdot SL}) / (e^{k' \cdot t} + e^{k' \cdot SL} - 1) + 1)$	$D(T) = \ln(10)/k$ $\text{Log } D(T) = \log D_{ref} - (T - T_{ref})/z$

$$\log D(T) = \log D_{ref} - \frac{T - T_{ref}}{z} \tag{15}$$

Table 1 summarizes the model equations of these models.

2.3.2 Inactivation Models for Dynamic Temperature Conditions

The description of microbial inactivation under dynamic conditions (i.e., varying temperature) is more complex than for isothermal treatments. Each dynamic experiment covers a range of temperatures, so it is not possible to isolate the effect of each temperature through experimental design. Regardless, a conceptual separation between primary and secondary models is still useful for model definition. It also eases the comparison between models built based on data gathered under isothermal and dynamic conditions. Therefore, we recommend following the two-step approach for model definition also for dynamic models. Another source of complexity for models under dynamic conditions is that they must be defined using differential equations. For most cases, these models do not have an analytical solution, so they must be solved using numerical methods [32].

Primary models for dynamic microbial inactivation can be constructed by calculating first derivatives with respect to time for the primary models defined for isothermal conditions. Regarding secondary models, the same algebraic equations can often be used for isothermal or dynamic conditions. Table 2 illustrates the primary and secondary microbial inactivation models constructed based on the isothermal models described in Table 1.

It has been reported in several scientific studies that dynamic heat treatments give rise to phenomena that cannot be observed under isothermal conditions. An example of this is stress acclimation, which has been observed in numerous empirical studies [33–39]. In industrial settings, thermal treatments require heating up the ingredients (usually from room temperature; <35 °C) to the treatment temperature (e.g., ~72 °C for pasteurization). Empirical studies have shown that the heating rate can affect the stress resistance of vegetative cells, with lower heating rates promoting higher

Table 2
Common models for describing microbial inactivation under dynamic conditions

	Primary model	Secondary model
Bigelow	$d \log N/dt = -\ln(10)/D(T)$	$\text{Log } D(T) = \log D_{ref} - (T - T_{ref})/z$
Peleg	$d \log N/dt = -b(T) \cdot n \cdot (-\log S/b(T))$^$((n-1)/n)$	$b(T) = \ln(1 + \exp.(k_b(T - T_c)))$
Mafart	$d \log N/dt = -p(1/\delta(T))$^$p \cdot t$^$(p-1)$	$\text{Log } \delta(T) = \log \delta_{ref} - (T - T_{ref})/z$
Geeraerd	$dN/dt = 1/(1 + C(0)) \cdot k' \cdot (1 - N_{res}/N(t)) \cdot N(t)$	$D(T) = \ln(10)/k$ $\text{Log } D(T) = \log D_{ref} - (T - T_{ref})/z$

resistance [33–39]. It has been hypothesized that the reason for this increased resistance is a physiological response of the microbial cells induced by the application of sublethal temperatures. Because stress acclimation can only be observed under dynamic conditions (in isothermal treatments, heating can be considered instantaneous), it is impossible that mathematical models developed based on data gathered under isothermal conditions could include this effect (empirical models cannot account for information not included in the data). This is reflected in model predictions underestimating the microbial count, with larger bias for lower heating rates, when stress acclimation is relevant [40]. Therefore, the description of microbial inactivation under dynamic conditions may require model hypotheses more complex than those used to describe isothermal inactivation.

The study of microbial inactivation under dynamic conditions is relatively novel compared to isothermal conditions. Plenty of empirical studies have shown deviations with respect to the inactivation expected based on isothermal experiments, both for high and low inactivation rates [34, 35, 38–44], some of them also proposing mathematical models. However, the lack of a detailed mechanistic explanation for these deviations makes it hard to define a clear protocol to define models for dynamic conditions. As general guidelines, we recommend using as baseline classical models derived from isothermal ones (Table 2). This allows checking for any potential effect by the dynamic treatment (e.g., increase or reduction in resistance) and how it can be affected by processing parameters (e.g., if the experimental design includes various experiments with different heating rates). If the deviations show a clear pattern (e.g., lower heating rates increase bias), the model equations can be expanded with novel hypotheses. Note that these equations should be suitable for the novel hypotheses. For instance, if the hypothesis of the study is that the history of the cells affects stress resistance (e.g., stress acclimation), the model should be written using differential and integral equations, not algebraic equations. An illustration of this sequential approach, based on the expansion of classical isothermal models, can be found in Garre et al. [40].

2.3.3 The Effect of Other Environmental Factors

Unlike for temperature, where the effect of changes in temperature in the inactivation rate are relatively well known (although there is still ongoing research for dynamic treatments), there are many knowledge gaps regarding the effect of other environmental factors on microbial kinetics, especially regarding interactions between environmental factors. Nevertheless, the model equations presented in Table 1 can usually be adapted to describe microbial inactivation due to other agents. Regardless of the type of stress, survivor curves are usually log-linear, have a slight curvature, or are sigmoidal. Hence, the primary models used for thermal inactivation are usually applicable to other types of stress. On the other hand, secondary models need to be adapted. Although various approaches can be found in the literature, it is common to use a polynomial relationship for the logarithm of the inactivation (or, equivalently, the D-value). For instance, Alvarenga et al. [45] used a polynomial relationship to describe the effect of temperature, pH, and Brix content in the inactivation of heat resistant molds. A similar approach based on existing primary models and polynomial relationships has also been used for high hydrostatic pressure [46]. Nonetheless, this field has relatively large knowledge gaps. For that reason, it is strongly advised to follow the protocol proposed in this chapter, making an emphasis in model validation to compare and select among the various candidate models.

2.4 Experimental Design

The last step of the conceptual phase is the definition of an experimental setting suitable for answering the scientific questions posed in the scope, potentially discriminating between the different candidate models. This step can be divided into two smaller parts. The first one is a conceptual one, where the general variables to include in the experimental design and their range are defined. Although there are several ways to approach this task, we suggest an approach similar to the one illustrated in Table 3 (for a theoretical *Listeria monocytogenes* study) to link the scope of the study with the experimental design.

The second part is the detailed definition of the experimental design. This step consists in a detailed description, including the number of repetitions, number and location of the time points, and the specific treatment temperatures. The main goal of this step is to generate a dataset with enough information to calculate parameter estimates with the required level of precision. It is thus inherently different to the other part of the design, whose main goal is to define a study broad enough to answer the scientific questions posed in the scope. It is certainly not simple to predict the amount of data required to estimate the model parameters with enough precision before actually doing the experiments (although it can be estimated based on numerical simulations) [20]. Nonetheless, their precision will depend mostly on three factors:

Table 3
Example of the conceptual phase of experimental design

Scope	Experimental design
What are the inactivation kinetics of *listeria monocytogenes*	The experiment will be performed using several *L. monocytogenes* strains
In milk	Milk will be obtained by a local supplier
Under isothermal conditions	Experiments will be carried out in equipment suitable to study microbial inactivation under isothermal conditions
At temperatures below pasteurization	Experiments will be carried out between 60 and 70 °C
And how are they affected by incubation conditions?	Cells will be precultured at different temperatures

- *The number and position of the time points* will mainly define the precision of the parameters of the primary model. Nonetheless, it is important to notice that deviations in the primary model parameters are carried out to the secondary models. As a general rule, additional time points will result in lower uncertainty. Nevertheless, several studies have shown that the application of Optimal Experiment Design (OED) can increase the precision of the parameter estimates without increasing the experimental effort [23, 47, 48]. Although OED requires a precise knowledge of the parameter estimates, results from OED can be used as guidelines for experimental design [20, 23]. Apart from these recommendations based on OED, it is advised that each repetition of the experiment includes at (the very) least five time points to at least observe any deviation from nonlinearity. Moreover, the difference between the initial and the final microbial count should be higher than the experimental error (i.e., at least two log CFU/g). Regardless of their number or location, the number of dilutions and the plated volume for each time point should be planned expecting a microbial count between 30 and 300 colonies per plate (lower counts would result in a deviation from normality, whereas higher counts may involve interactions between individual colonies). Although this may be hard to do before actually doing the experiment, the literature review can provide preliminary estimates of the model parameters that can predict the expected microbial count for each time point (e.g., using *bioinactivation*), aiding in experimental design.
- *The number and range of the different conditions tested* (e.g., number of temperatures) will determine the precision of the secondary model. This step of the experimental design has two dimensions: the density of the observations (e.g., how many

temperatures are tested) and the range of the conditions (e.g., the temperature range). The former will determine the parameter uncertainty of the secondary model, so an inadequate design is often easily identified in the validation step. Limitations in the latter are more nuanced because predictive microbiology uses empirical models. Therefore, it is not possible to ensure that the simple relationships (e.g., log-linear) fitted to the empirical data are held outside of the observed range. For that reason, these models should be used with extreme caution for conditions outside of those analyzed experimentally (e.g., higher temperatures). Therefore, it is advisable that the experimental range should envelope the actual conditions to be simulated. However, this is not always possible for microbial inactivation, as pasteurization temperatures (~70 °C) usually result in inactivation rates too high to observe survivor curves with enough time points [49]. Nevertheless, this limitation should be reported and should be carefully considered when validating and using the model.

- *The number of repetitions* of the experiment will also define the precision of the parameter estimates. From a statistical point of view, most fitting algorithms consider each time point as an individual observation [50], so a single repetition with enough time points would be enough. However, models in predictive microbiology describe a biological system, where variability is usually relevant [51]. Moreover, it is impossible to ensure that there were no errors during sample preparation [52]. Consequently, every experiment should be repeated at the very least twice. Regardless of the number of repetitions, it must be reported whether experiments are biological replicates (using independent cultures) or technical replicates (with the same culture), as this will influence the variation in the observations (due to varying sources of variability). Considering that the standard error of the parameter estimates is a reflection of variability and uncertainty [53], these estimates will be affected by whether the experimental design consists of technical or biological replicates. For that reason, it is crucial for a correct interpretation of uncertainty and variability to report the type of replicate. Furthermore, if the purpose of the study is to describe variability in the microbial response, the experimental design should include a relevant number of biological replicates at the relevant level (i.e., biological replicates of the same bacterial strain or from different strains).

2.5 Data Gathering

The next step in the protocol is the first one of the practical phase. It consists in gathering the experimental data that will be used to build the model. This can be divided into three steps: preparation of bacterial media and bacterial cultures, preparation of the heating device, and obtention of time points and counting (*see* **Note 2**).

2.5.1 Preparation of Media and Bacterial Cultures

Nowadays, most bacterial cultures are kept in frozen stocks at −80 °C, a condition that is dramatically different to the one bacterial cells encounter in the food chain. For this reason, before performing microbial inactivation experiments, bacterial cultures have to be reconstituted. Hence, the first medium required in an inactivation experiment is the one used to grow the bacterial cultures, typically an optimal growth medium. It can be necessary to perform two or more passes (subculturing from a small sample of the culture, once it reaches stationary phase) before the bacteria are considered to be prepared to conduct inactivation experiments. The second medium is the one actually used to perform the inactivation experiments. It can vary from buffered peptone water to complex foods. Finally, a different medium (selective, liquid or solid, etc.) may be used to incubate the samples and estimate the number of surviving cells. Specific culture preparation conditions will depend on the particular bacterial species analized, so we recommend checking specific guidelines for each study (*see* Chapter 2).

Several reports described that the growth conditions can have a strong influence on the bacterial response to thermal treatments. For instance, Crespo Tapia et al. [54] reported an increase in the thermal resistance of *Listeria monocytogenes* populations grown in the presence of lactose. The impact of the growth conditions on the thermal resistance of the microorganisms is not limited to the physicochemical properties of the media. As an example, it has been reported that cells of *Lactobacillus plantarum* in the exponential growth phase have lower thermal resistance than cells in the stationary growth phase [55]. Therefore, the growth conditions must be selected carefully according to the goals of the study and reported in detail. In studies with an applied goal, it may be reasonable to aim for worst-case scenario conditions, i.e., cells grown in rich laboratory media until the early stationary phase. However, it is hard to define that any scenario is a “worst-case”. For instance, the same acid shock can increase the thermal resistance of one bacterial population and have the opposite effect on a different strain [56]. Hence, it is recommended to use “standard” conditions (growth media and incubation conditions) that would ease the comparison with data already existing in the literature or conditions that mimic those in the actual food chain.

The heating media will depend on the scope of the study and the experimental design. For studies with a fundamental scope, laboratory media are usually more suitable, as their results are generally more reproducible than with food products. However, food products are complex physicochemical systems, whose attributes can affect microbial inactivation (*see* Chapter 3) [57]. Hence, studies whose goal is to develop a predictive model for process design or risk assessment should also include experiments using the food product of interest. Regarding more practical matters, if the heating substrate is a lab medium, following the recommendations by the manufacturer usually provides reproducible results. When the heating medium is a food product, one of the most relevant questions is whether a pretreatment is needed to eliminate the background microflora. If the initial microbiota of the product is high, some of these cells may survive the heat treatment and form colonies in the recovery medium that may be indistinguishable from the ones by the pathogen of interest. Moreover, although interactions are usually less relevant for inactivation than for microbial growth, they can still influence the microbial response. Consequently, when using food products as heating media, it is advisable to measure the background microbiota of the untreated product (e.g., measuring the total aerobic bacterial counts). If its concentration is high, applying a pretreatment to the substrate to eliminate this biological contamination should be considered, or to use selective media or fluorescence staining for recovery/enumeration [58].

The recovery media should be selected according to the pathogen studied. When possible, ISO guidelines (or similar) should be used, as this will improve the reproducibility of the results. External microbial contamination and recontamination should be avoided, as discussed above. Additionally, the use of suboptimal/limiting media besides standard conditions can provide additional insight about the bacterial response. In principle, suboptimal recovery conditions (e.g., with concentrations of NaCl) would act as a hurdle that inhibits the growth of those microbial cells with sublethal damage. Then, the comparison between the cell counts obtained with optimal and selective media can be an indication of the level of sublethal damage in the population [56, 59].

2.5.2 Preparation of the Heating Device

Any error in the temperature will have a strong influence in the observed microbial response. As an illustration, for an isothermal experiment following the Bigelow model, any error in the temperature of the device (ΔT) will result in an error in the D-value of $\varepsilon_D = 10^{-\Delta T/z}$. Hence, according to this model and considering a z-value of 5 °C, an error of −1 °C in the temperature of the equipment will result in an observed D-value 58% higher than the "true" D-value at temperature T. Consequently, it is crucial for the quality of the experimental data to set up the heating device properly, minimizing errors in the actual temperature applied to the microorganism.

Specific recommendations vary depending on the equipment used to perform the experiment. As a general rule, the temperature on the heating device should be stable before the heat treatment starts. For instance, in the case of isothermal treatments, the device should have stabilized at the treatment temperature before the bacterial cultures are inoculated/submerged. For dynamic treatments, the temperature of the device should be stable at the initial treatment temperature.

Regardless of the equipment used, we strongly advise to measure the temperature directly in the heating menstruum as a verification. Most thermoresistometers record the temperature in the heating chamber through the duration of the experiment [60]. This data should always be checked looking for any unexpected deviations. In the case of tubes in a water bath, it is not practical to measure the temperature in every single test tube. Although this information can be informative, the temperature of the water bath may not be representative of the temperature in the heating media because the wall of the test tube will introduce a temperature gradient between the water bath and the heating menstruum. Consequently, we recommend using some uninoculated test tubes as temperature controls. It is important to realize that, for the microorganism, the treatment temperature is not the one of the water bath but the one of the heating menstruum. Therefore, the models should be developed based on this temperature when possible.

Besides that, it is possible that some bacterial cells may stay in the environment and contaminate the surroundings of the heating device (e.g., the vessel of a thermoresistometer or the sampling area). For this reason, the device should be cleaned and disinfected regularly, minimizing the probability of cross contamination between experiments.

2.5.3 Obtention of the Time Points and Counting

The exact procedure followed to obtain the samples will depend on the heating device. In the case of test tubes in the water bath, each tube usually corresponds to one time point. On the other hand, if the experiment is done in a thermoresistometer, liquid samples will be taken from the heating vessel at preset times. Regardless of this, it is crucial that samples are cooled down immediately (usually by immersing the tubes in cooled ice or, if they are small volumes, diluting them in liquid medium such as peptone water at room temperature). Otherwise, the temperature in the heating media will still be high, resulting in additional microbial inactivation and a consequent overestimation of the effectiveness of the treatment (e.g., underestimation of the D-value).

As a general rule, dilutions should be plated in duplicate for greater accuracy and incubated at conditions favorable for the growth of the microorganism of interest (following international standards when possible). After incubation, the number of colonies on each plate can be counted by hand or using an automatic colony counter. Then, the microbial count (N) in CFU/ml can be calculated as shown in Eq. (16), where C stands for the number of cells counted in the plate, V_{plate} is the volume of heating menstruum plated, f is the dilution factor (10 for decimal dilutions), and d is the number of dilutions.

$$N[\text{CFU/ml}] = C[\text{cells}]\tilde{n}\frac{1}{V\text{plate}[\text{ml}]}\tilde{n}f^{d} \tag{16}$$

Regardless of the method used, the number of colonies should be converted to "log-counts" (in log CFU/ml) before modeling. The main reason for this is that the regression model makes the hypothesis that the variance of the model residuals is constant (homoscedasticity). In most situations, the variance of the microbial count is proportional to its expected value, so a log-transformation shall stabilize the variance. The statistical grounds for this are further described in Subheadings 2.6 and 2.7.

2.6 Parameter Estimation

2.6.1 Model Fitting Under Isothermal Conditions

Due to the separation between primary and secondary models, model fitting of isothermal experiments in predictive microbiology can be approached in two manners. The traditional one is to fit first the primary model for each isothermal experiment individually, resulting in a dataset of parameter estimates (e.g., D-values) calculated at each temperature. On a second step, the secondary model is fitted to this dataset of parameter estimates. Due to the sequential nature of this method, it is usually named "two-step approach." As an alternative, one can fit both the primary and secondary model to the whole dataset using nonlinear regression. This method is usually called the "one-step approach."

The one-step approach has several advantages from the point of view of the accuracy of the parameter estimates, usually providing more reliable estimates with lower uncertainty [61, 62]. On the other hand, the two-step approach explicitly provides information of the intermediate calculations (e.g., the D-values estimated at each temperature), making it easier to validate the fitness of the secondary model. For that reason, it is usually a good idea to consider them as complementary and use them both in the analysis (*see* **Note 3**).

2.6.1.1 Model Fitting of Isothermal Models Following a Two-Step Approach Using the Web Version of *Bioinactivation*

The web version of *bioinactivation* includes a module to fit isothermal models following a two-step approach. The first step when using this application is to upload the inactivation data. It must be formatted as a table with three columns: the elapsed time, the treatment temperature, and the microbial count. It can be input as an Excel file or a text file or manually within the tool (Fig. 2). Once the data has been loaded, the application shows a plot of the data, which can be used to check if the data was uploaded correctly.

Once the data has been uploaded, the first step in parameter estimation is fitting one primary model per treatment temperature. The application allows the user to select any primary model among those shown in Table 1 (Fig. 3). If the fitting was successful, the application generates a plot comparing the fitted models against the experimental data for each temperature (Fig. 3).

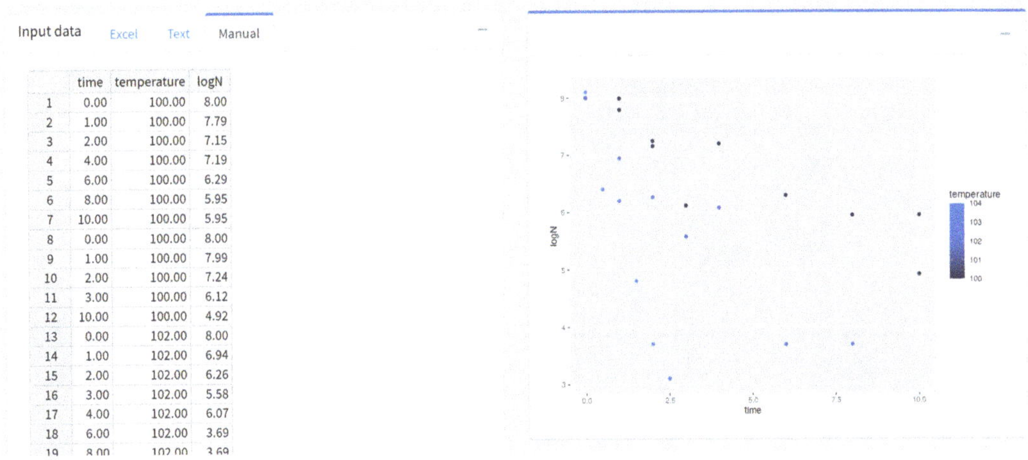

	time	temperature	logN
1	0.00	100.00	8.00
2	1.00	100.00	7.79
3	2.00	100.00	7.15
4	4.00	100.00	7.19
5	6.00	100.00	6.29
6	8.00	100.00	5.95
7	10.00	100.00	5.95
8	0.00	100.00	8.00
9	1.00	100.00	7.99
10	2.00	100.00	7.24
11	3.00	100.00	6.12
12	10.00	100.00	4.92
13	0.00	102.00	8.00
14	1.00	102.00	6.94
15	2.00	102.00	6.26
16	3.00	102.00	5.58
17	4.00	102.00	6.07
18	6.00	102.00	3.69
19	8.00	102.00	3.69

Fig. 2 Illustration of the data input for fitting isothermal inactivation models using a two-step approach in *bioinactivation*

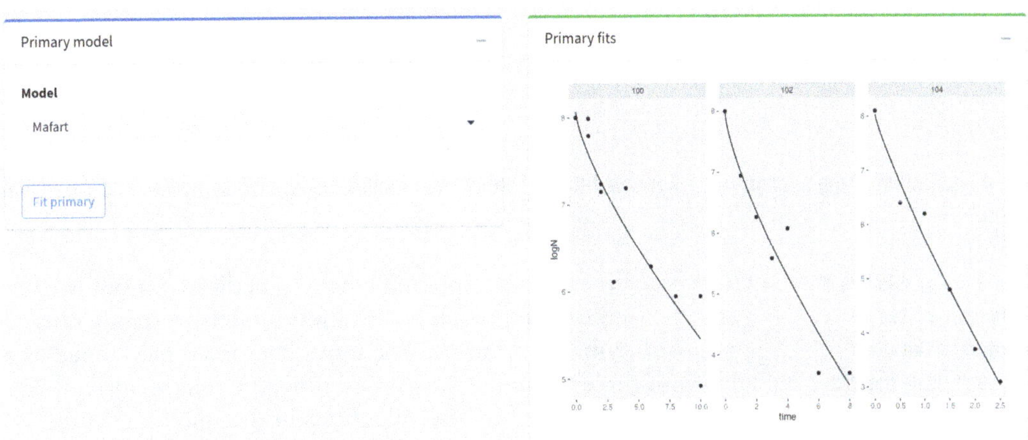

Fig. 3 Illustration of the model selection and primary model fits reported by *bioinactivation* for model fitting of isothermal models using a two-step approach

The second step for parameter estimation is fitting the secondary model to the dataset of parameter estimates calculated in the first step. The application shows a table with these values (Fig. 4). After selecting the reference temperature, clicking the "Fit secondary" button fits the secondary model to the data. Note that, for Weibullian models, the application does not consider a unique shape factor (p in Mafart model, n in Peleg model) for every temperature, unlike usually recommended [28]. Currently, this can only be done in the web version of *bioinactivation* using the module for one-step fitting.

If the fitting was successful, the application generates a plot comparing the fitted secondary model against the parameter estimates of the primary model, as well as statistical information of the fit (Fig. 4).

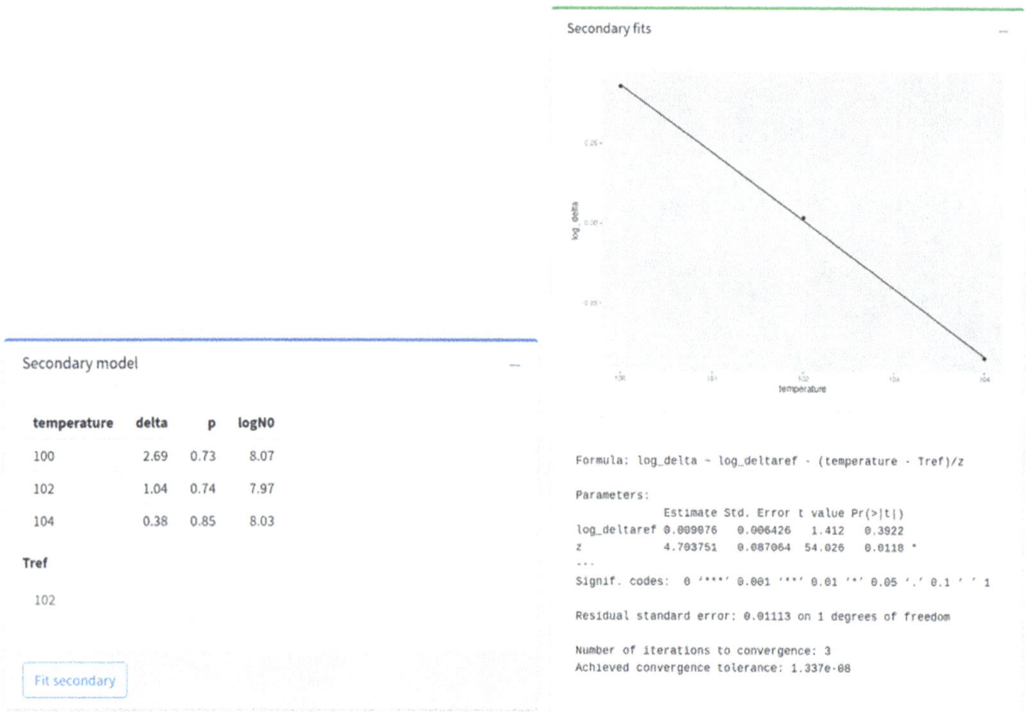

Fig. 4 Illustration of the second step for model fitting of isothermal models using *bioinactivation* and the model output

2.6.1.2 Model Fitting of Isothermal Models Following a One-Step Approach Using the Web Version of Bioinactivation

The *bioinactivation* web app implements a dedicated module to fit using the one-step approach inactivation models to data gathered under isothermal conditions. The input data must be defined as a table with three columns: the elapsed time, the treatment temperature, and the number of log-reductions in the microbial count. It can be uploaded as an Excel file or a text file or manually (Fig. 5). Once the data has been loaded, the application creates a plot that can be used to check if the data was interpreted correctly. Note that, unlike for the two-step fitting, this module requires the user to input the logarithm of the number of reductions ($\log S = \log N - \log N_0$), not the microbial count (*see* **Note 4**).

The application allows fitting most models defined in Table 1 (Fig. 6). The Geeraerd [30] model cannot be used in this case because the one-step approach requires the definition of secondary models for every parameter of the primary model, but parameters SL and N_{res} do not have a broadly accepted secondary model. The fitting is based on nonlinear regression, so the user must define starting values for every model parameter. This can be done based on the graphical representation of the data provided by the tool (Fig. 5). Moreover, the tool can fix any model parameter prior to the fit. This may be useful to improve parameter identifiability or to explore different models.

Input data Excel Text Manual

	time	temperature	logS
1	0.00	100.00	0.00
2	1.00	100.00	-0.21
3	2.00	100.00	-0.85
4	4.00	100.00	-0.81
5	6.00	100.00	-1.71
6	8.00	100.00	-2.05
7	10.00	100.00	-2.05
8	0.00	100.00	0.00
9	1.00	100.00	-0.01
10	2.00	100.00	-0.76
11	3.00	100.00	-1.88
12	10.00	100.00	-3.08
13	0.00	102.00	0.00
14	1.00	102.00	-1.06
15	2.00	102.00	-1.74
16	3.00	102.00	-2.42
17	4.00	102.00	-1.93
18	6.00	102.00	-4.31
19	8.00	102.00	-4.31
20	0.00	104.00	0.00

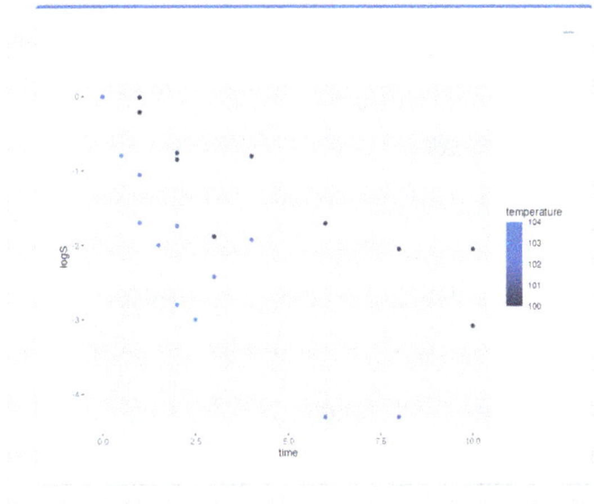

Fig. 5 Illustration of the data input for fitting isothermal inactivation models using a one-step approach in *bioinactivation*

If the fitting was successful, the application creates a plot comparing the fitted models against the experimental data (Fig. 7). This can be used to detect possible convergence issues of the algorithm or inadequacy of the model hypotheses (i.e., when there is a clear deviation between the model predictions and the data points). Furthermore, the application provides the statistical information of the fit (parameter estimates, standard errors, and error indexes), as well as several plots to evaluate the goodness of the fit based on the model residuals (see Subheading 2.7 for the statistical basis of these diagnostics).

2.6.2 Model Fitting Under Dynamic Conditions

Modeling microbial inactivation under dynamic conditions involves higher mathematical complexity, requiring solving and fitting models based on differential equations. For most cases, these models do not have an analytical solution, so they must be approximated using numerical algorithms. There is a large variety of algorithms to solve ordinary differential equations, and evaluating their convergence may not be simple in some situations. For that reason, it is advisable to use software packages that already implement modern algorithms (including convergence analysis), such as the *deSolve* package for R [63] or *SciPy* in Python [64]. As an alternative, one can use specific software tools to predict microbial inactivation under dynamic conditions that already implement the solution of the differential equations (ideally based on these modern algorithms), such as *bioinactivation*. In cases where these published tools are not used and the numerical algorithm is implemented de novo, the details of the implementation should be discussed in scientific publications, ideally sharing the code.

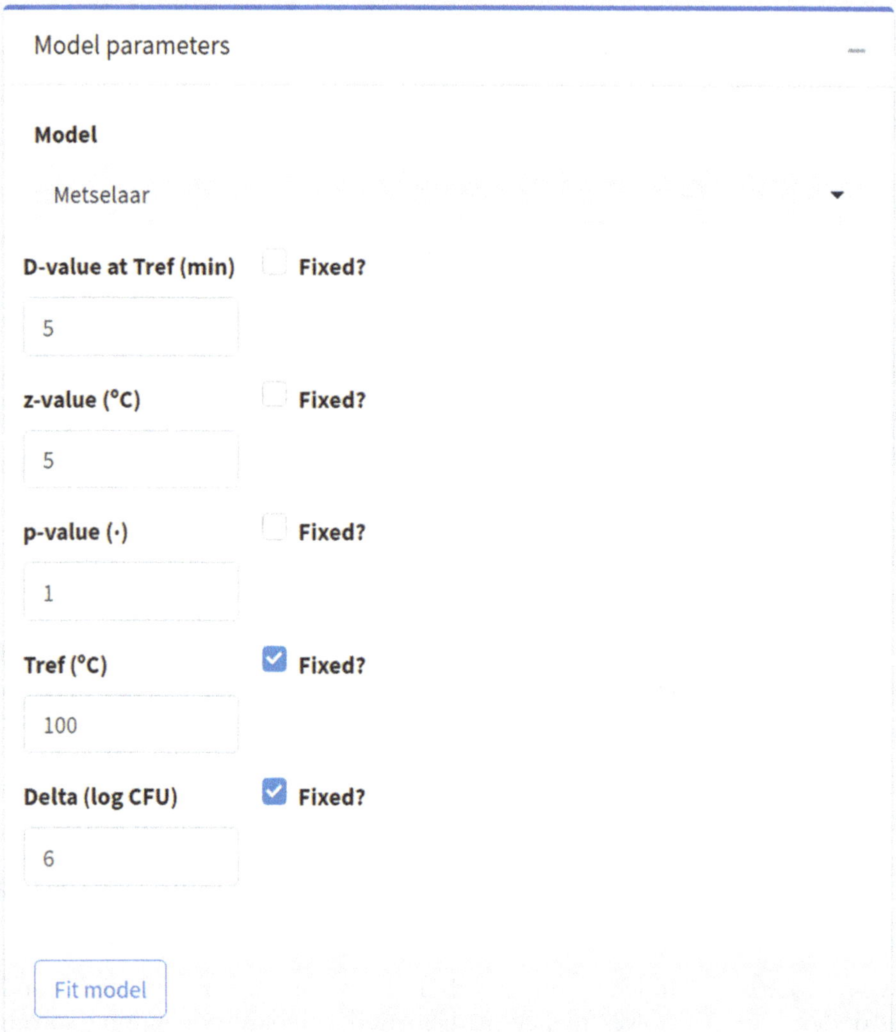

Fig. 6 Model selection and definition of model parameters to fit and their initial values for the one-step fitting in *bioinactivation*

Fitting models based on differential equations usually involves additional identifiability issues [47]. Therefore, fitting dynamic models often requires more complex algorithms, such as the Levenberg-Marquardt algorithm [65] or algorithms based on Monte Carlo simulations [66]. The implementation of these algorithms is certainly not simple, so the use of dedicated software packages for parameter estimation of dynamic models is again strongly advised, such as the *FME* package for R [67] or the AMIGO2 toolbox for MATLAB [68] or specific tools for predictive microbiology such as *bioinactivation*.

Fig. 7 Model output for isothermal models fitted using the one-step approach in *bioinactivation*

2.6.2.1 Model Fitting Dynamic Models Using the Bioinactivation Web Application

The online version of *bioinactivation* includes a module to fit inactivation models to data gathered under dynamic conditions. The data must be input in two different tables: one describing the temperature profile and a second one for the observed microbial count. The temperature profile can be uploaded as an Excel file or a text file or manually, as a table with two columns with the elapsed time and the treatment temperature at each time point (Fig. 7). Note that, for time points not included in the table, the tool estimates the temperature by linear interpolation (Fig. 8). This allows the definition of both ideal profiles (e.g., biphasic profiles) and profiles based on actual temperature measurements.

The microbial counts are input as a table with two columns: the elapsed time and the observed log-microbial count (Fig. 8). They can also be uploaded as an Excel file or a text file or manually.

The inactivation model to fit can be selected among those included in Table 2 using a dropdown box (Fig. 9). The tool includes two algorithms for model fitting, both of which require

Fig. 8 Input of the temperature profile and the microbial counts for fitting dynamic models in *bioinactivation*

initial values for every model parameter. In order to ease their definition, the tool depicts a plot comparing the experimental data against the prediction corresponding to the initial guess of the model parameters (black line; Fig. 9). In order to mitigate convergence issues, the tool defines upper and lower bounds for every model parameter. These are plotted in the same figure to ease the definition of realistic bounds (purple and blue lines; Fig. 9). Every data point should be contained within the area defined by these two lines before fitting the model. Furthermore, the tool allows fixing any model parameter before the fit. This may serve to mitigate parameter identifiability issues or to explore different types of models.

Besides the "traditional" nonlinear regression algorithm, *bioinactivation* also includes an adaptive Monte Carlo algorithm, which can be more robust than nonlinear regression for models based on differential equations [66]. Note that, unlike nonlinear regression, this algorithm does not include an automatic check of

Fig. 9 Definition of the inactivation model, as well as what parameters to fit, their initial values, and bounds for dynamic data in *bioinactivation*

convergence. Therefore, the user must choose the number of samples from the Markov chain. Furthermore, the tool allows setting the burnin length and the frequency at which the covariance matrix is updated for this algorithm. The reader is referred to the specific scientific literature for the interpretation of these terms [66, 67]. Finally, because the adaptive Monte Carlo algorithm is stochastic, the tool includes a button to reset the seed, providing reproducibility of the results.

If the fitting was successful, the tool generates a plot comparing the fitted model and the experimental data (Fig. 10). Furthermore, it creates a table reporting the parameter estimates and their confidence intervals (or credible intervals in case of the adaptive Monte Carlo algorithm) and a second table with various error indexes (Fig. 10). It also generates various plots that can be used to assess the goodness of fit based on the distribution of the residuals (see Subheading 2.7 for the statistical basis for these diagnostics).

2.6.3 Alternatives to Bioinactivation for Model Fitting of Inactivation Experiments

Although the protocol presented in this chapter has been supported on the use of *bioinactivation* for model fitting and prediction, some alternative tools are available for modeling microbial inactivation (*see* Table 2, Chapter 1) [19]. General-purpose applications (programming languages, applications for model fitting, FSK-ML) can also be used to fit the models described here. Their advantage is their great flexibility, allowing the implementation of practically any inactivation model. On the other hand, the user must define the models and prediction/fitting algorithms,

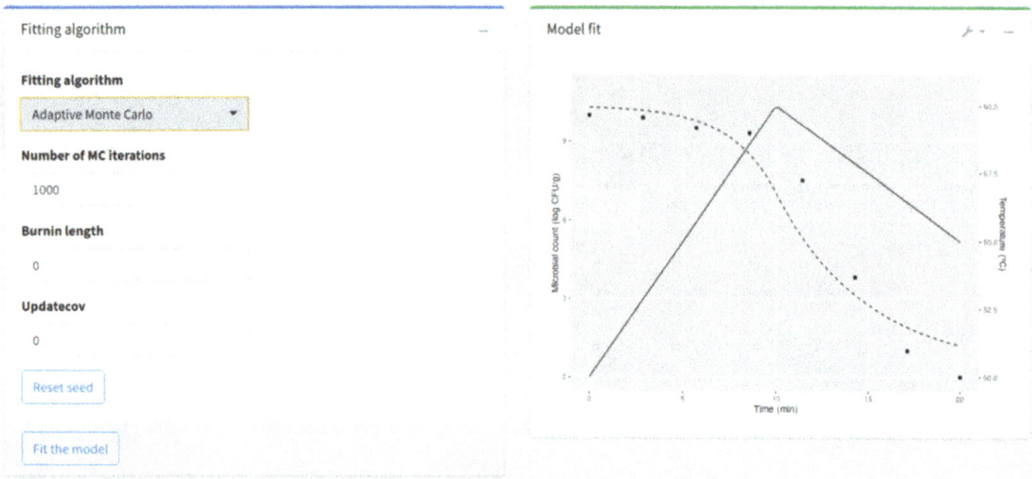

Parameter estimates

Parameter	Estimate	Std. Error
D_R	2.54	0.83
z	5.84	3.92

ME	RMSE	loglik	AIC	AICc	BIC	Af	Bf
-0.26	9.17	11.94	-19.88	-17.48	-19.72	1491364810.91	0.55

Fig. 10 Definition of the parameters of the fitting algorithm and model output for fits of dynamic inactivation models in *bioinactivation*

increasing the time required to prepare the analysis. Moreover, using this type of tool may reduce reproducibility, as the original code used for model fitting may be hard to share with other researchers.

2.7 Model Validation

The final step in the protocol is model validation. This is a crucial step for the development of empirical models, which should be performed with extreme care and a critical attitude. As illustrated in Fig. 1, model validation can identify relevant model deficiencies that may require a return to the conceptual phase. Although there are different approaches to validate an empirical model for microbial inactivation, we recommend the following steps:

1. Analysis of the model residuals.
2. Evaluation of the parameter and prediction uncertainty.
3. Comparison against already published data.
4. Comparison against independent data (when relevant/ applicable).

2.7.1 Analysis of the Model Residuals

In most cases, inactivation models are fitted to data using (non)-linear regression (or similar methods), which considers that the deviation between the model predictions and the observations are i.i.d random variables following a normal distribution with constant variance [69]. This is a crucial model hypothesis that must be verified to mitigate the possibility of introducing a bias in the parameter estimates and/or model predictions. Furthermore, deviations with respect to these hypotheses may be a valuable source of information, being an argument to use more complex mathematical models or designing new experiments. In this chapter, we focus on three deviations with respect to these hypotheses: deviations from normality, deviations from independence, and heteroskedasticity (non-constant variance). Although some statistical tests are available to evaluate these hypotheses, we recommend an analysis based on graphical representations, which we consider is usually more insightful. A more in-depth discussion of these topics can be found in the scientific literature [69–71].

Firstly, the normality of the residuals can be verified based on a histogram of the residuals overlapped with the probability density function of a normal distribution with mean zero and the standard error of the residuals ($SER = (\Sigma(y_{pred} - y_{obs})^2/(n-p))^{1/2}$). Figure 11a illustrates a histogram without any visible deviation with normality, whereas Fig. 11b shows a histogram with a clear deviation. This type of histogram can occur for microbial inactivation when a unique model has been fitted to a dataset including data from different conditions (e.g., different media). The particular way in which the histogram will deviate from the theoretical one will depend on several aspects of the data and the model. For instance, if a unique model is fitted to data that includes two well-separated categories (e.g., two different media with an effect on microbial inactivation), the histogram will show two peaks (as shown in Fig. 11b). If the data is balanced (e.g., if both media have the same number of data points), the two peaks will have the same height, whereas any imbalance will cause one peak to be higher than the other. Nevertheless, various model misspecifications can result in anomalous histogram of the residuals, so any deviation should be analyzed with care.

The second regression hypothesis that should be tested is the independence of the residuals. This can be done with a plot of the residual versus the predicted values. If the model hypotheses are correct, the plot should not show any obvious trend (Fig. 11c). As

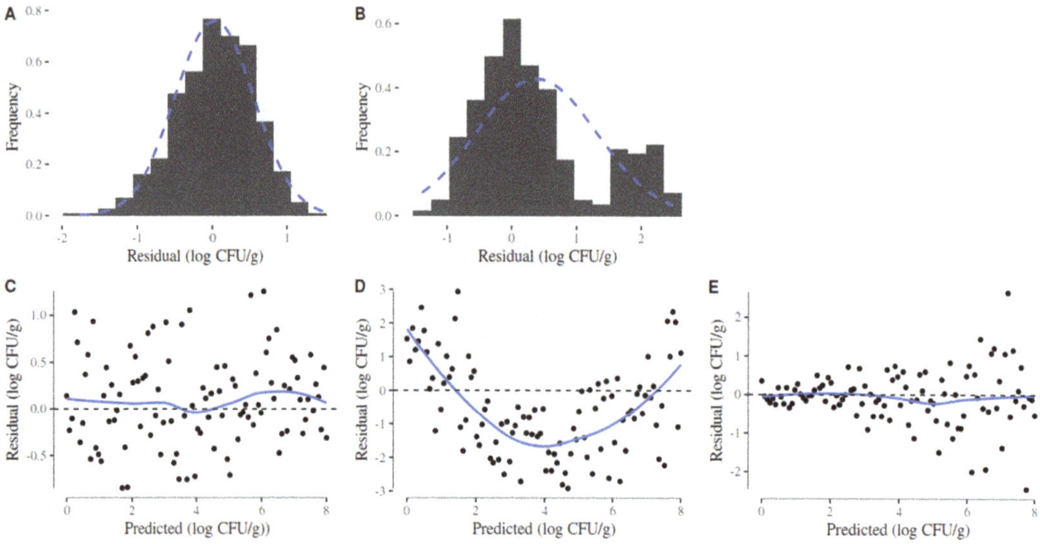

Fig. 11 Model diagnostics based on the distribution of the residuals. (**a**) Expected histogram of the residuals. (**b**) Typical anomalous histogram of the residuals. (**c**) Expected residual plot. (**d**) Typical residual plot for an underparameterized model. (**e**) Typical residual plot for heteroscedastic data

an example of a case violating this hypothesis, Fig. 11d shows a plot where the residuals have a "parabolic" trend, which is a typical result of fitting a linear model to nonlinear inactivation data. An alternative graphical representation that provides similar information is a predicted vs. observed plot.

The third regression hypothesis to be tested is that the variance of the residuals is constant (i.e., homoskedasticity). A plot of the residuals versus the predicted values is also the recommended tool to assess this hypothesis. As an illustration, Fig. 11e depicts a residual plot where the points show a "funnel"-like shape, with higher spread for higher predictions. For microbial inactivation studies, observations usually have higher variance when the microbial count is low [20, 72]. Therefore, a relatively small heteroskedasticity of the residuals is usually not an indicator of a deviation of the model hypotheses, rather a reflection of the experimental limitations. Although this deviation can be justified, it can introduce a bias in the parameter estimates. Hence, when possible, it is recommended to evaluate how different strategies affect the model fits [73]. Nonetheless, if the heteroskedasticity is severe or does not have the expected "shape" (i.e., variance is not higher for lower counts), its sources should be analyzed with care because they may introduce a significant bias in the model (Fig. 11) [74]. Alternatively, one can use a statistical method that includes these sources of heteroskedasticity in the likelihood functions, such as the Most Probable Curve method.

2.7.2 Evaluation of the Parameter and Prediction Uncertainty

Models in predictive microbiology are based on empirical data, so their parameter estimates and their predictions are affected by different sources of uncertainty, as well as the sources of variability (*see* **Note 5**) that may be of relevance [75]. This is usually quantified by the standard error of the parameter estimates or, equivalently, their confidence intervals. In some situations, the uncertainty of the model may be too high to draw any relevant conclusion based on the model predictions or the parameter estimates. Considering that uncertainty can be reduced by gathering additional experimental data [76], the assessment of whether additional data is needed is a relevant part of model validation [77]. Depending on the scope of the study, this assessment may be focused on the uncertainty of the parameter estimates, the uncertainty of the model predictions, or both.

Many studies (especially those with a fundamental scope) base their conclusions on the variation of the parameter estimates. For instance, the effectiveness of an antimicrobial can be evaluated by comparing the inactivation rates estimated in experiments with and without the antimicrobial [78]. If the experimental data was of insufficient quality or quantity, the parameter uncertainty would be too high to observe any significant difference, even when there is an effect (low statistical power). The conclusions of other studies (especially the ones with a more applied focus) are focused on the model predictions. Because they are based on model parameters affected by uncertainty, uncertainty should also be included in model predictions. Although there are several methods to include it, the simplest one is usually based on Monte Carlo simulations [18, 79]. If the quantiles (or equivalent measure) are deemed too high to draw any meaningful conclusion, the experimental design should be expanded. It is usually a good idea to use the parameter estimates (albeit imprecise) when improving the experimental design, as they can improve the selection of, for instance, time points. Moreover, it can be considered the application of OED at this step, as in some situations it can strongly reduce the experimental effort required or, at least, provide guidelines for experimental design [20, 23].

2.7.3 Comparison Against Already Published Data

Microbial inactivation has been studied using mostly the same model equations for several decades. Therefore, for most scenarios, it is possible to find previously published studies with a similar experimental design (e.g., same microbial species and similar media). Although these studies can be found using online repositories of scientific studies (Scopus, Google Scholar, etc.), this process may be cumbersome due to the large number of nonrelevant studies returned by these resources. For that reason, it is recommended to use some of the databases available online for predictive microbiology (*see* Table 2, Chapter 1).

The comparison between the results of different studies can be made in several ways, mostly depending on the format of the data and the type of model used in each study. In some cases, only microbial counts (or the number of reductions) for one or more treatment configurations are available. In this case, observations can be compared against the model predictions (including uncertainty) using, for instance, a t-test. However, this comparison only provides a limited view of the microbial response (e.g., it does not show if there are nonlinearities in the survivor curve). For that reason, when available, it is usually more informative to directly compare the models fitted to the data. This can be done in two manners. When the same model equations are used, the parameter estimates can be compared (e.g., the D- and z-values) with a t-test based on the parameter estimates and their standard errors. When this is not the case, the model predictions of either study can be compared against the microbial counts of the other one. Then, the residuals can be evaluated using the techniques described in the previous section for model validation based on the residuals.

Deviations in the microbial response are somewhat expected between different studies. Potential causes for deviations between studies include, but are not limited to, the use of different microbial strains, differences in preculture and/or recovery conditions, microbial interactions, the use of different equipment, or differences in the matrix (also relevant for standardized media). Therefore, the existence of deviations does not prove that either model is wrong. Nonetheless, these deviations should be reported and analyzed with objectivity because, besides indicating potential model issues, they can also be a good reason to elaborate additional studies in the future.

2.7.4 Comparison Against Independent Data (When Relevant/Applicable)

A different way to validate predictive models is to compare the model predictions against independent data obtained within the same study and not used for model fitting [80, 81]. A simple way to do this type of validation is as follows:

- Separate the data into a training (for model fitting) and a validation set.
- Fit the model to the training set.
- Compare the predictions of the fitted model against the observations in the validation set using the tools for residual analysis described in Subheading 2.7 of this chapter.

Conceptually, the idea of this approach is to ensure that the model will be able to describe data under conditions different to those tested. Although simple, this method has several conceptual caveats. The first one is related to the division of the data in training and validation sets, which may introduce an element of subjectivity. To illustrate this limitation, let's use the example of inactivation under isothermal conditions where four temperatures are tested (60, 61, 62, and 63 °C). A possible division would be based on the treatment temperature, using some of the experiments for training and the rest for validation. This is a source of subjectivity, as it is not clear whether one should use two or three temperatures for validation and the rest for building the model. Furthermore, the particular temperatures to use for training are open to interpretation. If the training set includes two temperatures, should we use the two extremes or the two middle temperatures? Considering that different compositions of the training set will result in different values for the parameter estimates, this is an element of subjectivity potentially relevant for parameter estimates and model predictions.

Moreover, when designing the training and validation set, it is important to keep in mind that this method is intended to ensure that the model can predict conditions not tested experimentally. Therefore, the validation set should be different enough from the data used to fit the model. As an illustrative example for microbial inactivation, all the data points obtained at one temperature can be kept in the validation set. However, plenty of experimental studies have evidenced a log-linear relationship between the D-value and temperature. Therefore, it is very unlikely that a model trained using data at 60, 61, and 63 °C fails to predict data obtained at 62 °C. Consequently, this particular validation method would provide little to no information in this case.

These caveats can be somewhat mitigated by using more complex validation techniques. For instance, cross-validation makes several groupings of the data based on random sampling, repeating the process described above several times [82, 83]. Although this method involves a large increase in computation time, it omits the need for a manual split between validation and training sets. Nevertheless, many conceptual limitations of model validation remain when using cross-validation. Every experimental design will introduce some sort of bias (usually called "study effects" in the context of meta-analyses) [84]. If the validation data was generated within the same laboratory, it will contain the same study effects. Hence, this type of validation would not ensure the model would be able to describe the data generated under different laboratory conditions and, more relevantly, in the actual food chain.

Furthermore, it must be considered that the range of validity of predictive models for microbial inactivation is a complex scientific problem that, in the opinion of the authors, has not yet been solved (*see* **Note 6**). It is common to use data gathered under dynamic conditions to validate models fitted to isothermal data. However, even if the distribution of the residuals did not show any model deficiency, this does not imply the model can describe microbial inactivation under every dynamic condition. For instance, Garre et al. [48] compared the predictions of a model fitted under isothermal conditions against data gathered under dynamic conditions. For treatments with moderately high heating rates, they observed low deviations, which would indicate that the model is validated and can be used to describe dynamic inactivation. However, when the model predictions were compared against data obtained at low heating rates (<5 °C), they observed a relevant bias. Considering that similar conclusions have been reported in similar studies [39, 44], we can conclude that the fact that a model has been validated using one set of dynamic data does not imply that it will be universally valid for describing microbial inactivation under dynamic conditions. A similar logic can be applied for differences between different bacterial strains. Phenotypes can vary largely between strains of the same species [51, 85]. Therefore, the fact that a model trained using data from one bacterial strain and validated using data from a different one does not imply it is valid to describe every strain of that species.

These conceptual issues related to the range of validity of empirical models are, in our opinion, still an open scientific question. Consequently, when using a model in a particular case study (e.g., shelf life estimation), it is recommended to perform an additional validation with the relevant product (and, if possible, using specific bacterial isolates). And, even when there is a good agreement between the data and the model predictions, it is advised to account for the model uncertainty and the potential variability not included in the validation experiments.

3 Materials

3.1 Equipment to Establish Heat Inactivation Kinetics

Indirect Heating Methods They are based on a cell suspension introduced in a container of different sizes and materials.

- *Glass test/capillary tubes of different volumes have been used extensively. They were sealed by heat or other methods and then submerged in baths containing distilled water, mineral oils, etc. heated at the target temperature.*
- *A variation of the previous ones were Kooiman tubes [86], which consisted on metallic stainless steel tubes with a septum to inoculate the microbial suspension. The tubes were submerged in a heated bath.*

Open Methods Erlenmeyer flask in a bath. In this case, an Erlenmeyer containing a liquid medium was heated to the target temperature, and the microorganisms were then inoculated. Samples were taken at time intervals. The maximum temperature in this approach is close to 100 °C.

Table 4
Advantages and disadvantages of each method to establish heat inactivation kinetics

	Advantages	Disadvantages
Thermoresistometer	Instantaneous heating and cooling of samples Precise timing Labour saving Simplicity of operation Labor costs and low costs of upkeep	Initial cost
Test tubes	Simple, inexpensive equipment needed Easily opened for subculture	Slowest heat transfer Thermotolerance parameters are influenced by tube sizes Heating and cooling limit the temperatures that may be studied Accuracy Hazard of contamination during subculture
Capillary tubes	Requires lower volumes of highly concentrated suspensions comparing to test tubes Short come-up times Heated samples may be stored for later subculture or incubated unopened for observation	Laborious tasks of sealing and opening glass tubes Heating and cooling limit the temperatures that may be studied Accuracy Hazard of contamination during subculture
Erlenmeyer flask in bath	Short come-up times It requires only readily available equipment Heating and cooling lags are negligible when the method is properly employed	Confined to use of temperatures below the boiling point of water Extreme care is required to avoid splashing of the flask contents With some substrates, flocculation may occur

Thermoresistometers ne of the main problems associated with indirect heating methods at high temperatures is the time taken for the sample to reach the required temperature. One apparatus which has the advantage of short come-up times, uses small quantities of suspension, and does not suffer from the problems of calculating residence time distribution is the thermoresistometer [84]. The first thermoresistometer was designed by Stumbo for studying resistance of bacterial spores to temperatures above 115 °C (240 °F) (Stumbo, 1948). It used steam to heat the samples and had a maximum lag time of 0.3 s. Then, other equipment has been developed with the same aim.

(a) Thermoresistometer of Pflug *[87]. It was developed as an instrument designed to accomplish essentially the same objectives as the thermoresistometer of Stumbo. It was claimed that data could be obtained for temperatures up to 145 °C.*

(b) Thermoresistometer of Campden *[88]. In order to deal with and determine the heat resistance of bacterial spores, a computer-controlled thermoresistometer was designed and constructed. This thermoresistometer was based on the design of Pflug and Esselen [87]. The thermoresistometer was capable of operating between 100 and 150 °C (212–302 °F) with exposure times as short as 0.1 s. Five samples (0.01 mL) of spore suspension could be exposed to steam at each time/temperature combination using a computer-controlled piston-driven exposure system. Temperature control of the apparatus was achieved using platinum resistance probes linked to a computer, which then actuates a variable steam control valve.*

(c) Thermoresistometers Sala-Condon and Mastia. *The thermoresistometers Sala-Condon [89] and Mastia [60] consist of a laboratory-scale instrument to get heat resistance data of microorganisms under isothermal and nonisothermal treatments. The Sala-Condon allows to perform isothermal experiments with almost instant heating at temperatures up to 150 °C (302 °F). It is based on a stainless steel vessel with heating, shaking, and sampling systems [89]. The Mastia one is based on the one by Sala-Condon, with additional features. This device can mimic heat treatments and makes it possible to know the number of survivors at any time of the treatment; thus, it can be a helpful tool for food processors in order to estimate the level of safety of the treatments they apply. The thermoresistometer Mastia allows to perform thermal treatments in liquid media with a temperature profile that can be programmed within the maximum heating and cooling rates of the equipment (40 °C/min). The vessel of the thermoresistometer is constantly stirred during the treatment, ensuring a homogeneous temperature distribution (Table 4).*

4 Notes and Closing Comments

In this section of the chapter, we provide several comments that complement or complete various parts of the protocol presented in the previous section.

1. In the opinion of the authors, predictive modeling should not be seen as a final step of a streamlined process that applies model fitting to a dataset to create a mathematical model. If a predictive model is the goal of a study (or it will be used to base conclusions using inference), mathematical modeling should be considered already at the beginning of the study design. This would allow designing experiments that generate data that is more likely to be informative enough to build a model with

the required precision. For that reason, we emphasize in the protocol the consideration of the mathematical models in the conceptual phase of the study.

2. Regardless of the complexity of the model equations or the statistical algorithms used to analyze the model, empirical models will never contain information not included in the data. For that reason, it is crucial that the model is based on high-quality data. This involves conceptual elements, such as an appropriate experimental design (number of data points, number of conditions), as well as experimental ones (application standard microbiological protocols, number of repetitions). Consequently, when analysing, reporting and using a predictive model, it is important to consider the limitations of the data (potential experimental limitations, range of the conditions) as these will define some of the main limitations of the model.
3. Although current statistical software can fit highly complex models to data, it is important to consider whether that complexity is actually justified [21]. For that reason, we recommend beginning the analysis using models as simple as possible (i.e., include simple models in the list of candidate models). If the validation step shows a clear bias between these models and the data, the use of more complex models is justified. We believe this approach has several advantages. Firstly, it makes a strong statement for the need of the complex modeling approach, which may not be so clear if only the complex model is tested. Secondly, it makes it easier to identify potential limitations in previous studies that used more simple models for similar conditions.
4. Isothermal inactivation data can be fitted using either a one-step or a two-step approach. In our opinion, both approaches should be seen as complementary. The two-step approach provides valuable information about the relationship between the model parameters of the primary model and the environmental conditions that may be hidden in the one-step approach. Hence, it can be used as an exploratory step before applying the one-step approach to estimate the parameter values, as this approach usually estimates more realistic parameter values and uncertainties. For this, the use of software applications that include both approaches (e.g., *bioinactivation*) can be useful.
5. Although it has not been discussed in-depth in this protocol, (biological) variability is a topic of great relevance for shelf life estimation and risk assessment [51]. Biological differences between microbial populations can be responsible for differences in microbial kinetics. The relevance of variability can

depend strongly on the experimental design. For instance, Aryani et al. [90] observed that the differences in the inactivation kinetics between cells of different strains of *Listeria monocytogenes* were larger than between cells of the same strain. They also observed that the latter was higher than differences between technical replicates using the same strain. For this reason, it is crucial for a correct interpretation of the results to clearly state how many technical replicates were made for each experiment and whether these replicates were based on fresh cultures or if the same culture was used for all of them.

6. One last aspect to consider when using predictive models for microbial inactivation is that model validation still has many relevant limitations. Statistical methods for model validation are based on the data available, either generated as part of the study or gathered from previous scientific studies. As already discussed, it is not possible to describe with precision the range of validity of empirical models for microbial inactivation (e.g., the type of dynamic thermal profiles or the range of media). Moreover, biological variability is of high relevance for microbial inactivation. Hence, bacterial cells within the food chain are likely to have different inactivation kinetics than reference strains available in microbiology laboratories. Therefore, it is impossible that a model based on a reference strain will be able to describe with precision the inactivation kinetics of every wild strain, even if it describes the response of the reference strain without error. For these reasons, even if a model has been properly validated using the methods currently available, we should not expect that the model will be able to predict the actual bacterial response within the food chain without uncertainty.

References

1. Haase JK, Didelot X, Lecuit M et al (2014) The ubiquitous nature of L isteria monocytogenes clones: a large-scale multilocus sequence typing study. Environ Microbiol 16:405–416. https://doi.org/10.1111/1462-2920.12342
2. Bridier A, Sanchez-Vizuete P, Guilbaud M et al (2015) Biofilm-associated persistence of food-borne pathogens. Food Microbiol 45:167–178. https://doi.org/10.1016/j.fm.2014.04.015
3. Peng J, Tang J, Barrett DM et al (2017) Thermal pasteurization of ready-to-eat foods and vegetables: critical factors for process design and effects on quality. Crit Rev Food Sci Nutr 57:2970–2995. https://doi.org/10.1080/10408398.2015.1082126
4. Cebrián G, Mañas P, Condón S (2016) Comparative resistance of bacterial foodborne pathogens to non-thermal technologies for food preservation. Front Microbiol 7:1–17. https://doi.org/10.3389/fmicb.2016.00734
5. Van Asselt E, Zwietering M (2006) A systematic approach to determine global thermal inactivation parameters for various food pathogens. Int J Food Microbiol 107:73–82. https://doi.org/10.1016/j.ijfoodmicro.2005.08.014
6. Kourkoutas Y, Chorianopoulos N, Nisiotou A et al (2016) Application of innovative technologies for improved food quality and safety. Biomed Res Int 2016:1–2. https://doi.org/10.1155/2016/9160375

7. Krebbers B, Matser AM, Hoogerwerf SW et al (2003) Combined high-pressure and thermal treatments for processing of tomato puree: evaluation of microbial inactivation and quality parameters. Innov Food Sci Emerg Technol 4: 377–385. https://doi.org/10.1016/S1466-8564(03)00045-6
8. Duret S, Hoang H, Derens-Bertheau E et al (2019) Combining quantitative risk assessment of human health, food waste, and energy consumption: the next step in the development of the food cold chain? Risk Anal 39:906–925. https://doi.org/10.1111/risa.13199
9. Guillier L, Duret S, Hoang H-M et al (2016) Linking food waste prevention, energy consumption and microbial food safety: the next challenge of food policy? Curr Opin Food Sci 12:30–35. https://doi.org/10.1016/j.cofs.2016.06.006
10. Bigelow WD (1921) The logarithmic nature of thermal death time curves. J Infect Dis 29: 528–536
11. Bigelow WD, Esty JR (1920) The thermal death point in relation to time of typical thermophilic organisms. J Infect Dis 27(6):602–617
12. Esty JR, Meyer K (1922) The heat resistance of the spores of B. botulinus and allied anaerobes. J Infect Dis 31:650–664
13. McMeekin TA, Mellefont LA, Ross T (2007) Predictive microbiology: past, present and future. In: Modelling microorganisms in food. Elsevier, pp 7–21
14. Messens W, Hempen M, Koutsoumanis K (2018) Use of predictive modelling in recent work of the panel on biological hazards of the european food safety authority. Microb Risk Anal 10:37–43. https://doi.org/10.1016/j.mran.2018.04.002
15. van Lieverloo JHM, Bijlaart M, Wells-Bennik MHJ et al (2021) Thermal inactivation kinetics of seven genera of vegetative bacterial pathogens common to the food chain are similar after adjusting for effects of water activity, sugar content and pH. Microb Risk Anal 19: 100174. https://doi.org/10.1016/j.mran.2021.100174
16. Smelt JPPM, Brul S (2014) Thermal inactivation of microorganisms. Crit Rev Food Sci Nutr 54:1371–1385. https://doi.org/10.1080/10408398.2011.637645
17. Garre A, Clemente-Carazo M, Fernández PS et al (2018) Bioinactivation FE: a free web application for modelling isothermal and dynamic microbial inactivation. Food Res Int 112:353–360. https://doi.org/10.1016/j.foodres.2018.06.057
18. Garre A, Fernández PS, Lindqvist R, Egea JA (2017) Bioinactivation: software for modelling dynamic microbial inactivation. Food Res Int 93:66–74. https://doi.org/10.1016/j.foodres.2017.01.012
19. Possas A, Valero A, Pérez-Rodríguez F (2022) New software solutions for microbiological food safety assessment and management. Current Opinion in Food Science 44:100814. https://doi.org/10.1016/j.cofs.2022.100814
20. Garre A, Peñalver-Soto JL, Esnoz A et al (2019) On the use of in-silico simulations to support experimental design: A case study in microbial inactivation of foods. PLOS ONE 14. https://doi.org/10.1371/journal.pone.0220683
21. Zwietering MH (2009) Quantitative risk assessment: is more complex always better? Simple is not stupid and complex is not always more correct. Int J Food Microbiol 134:57–62. https://doi.org/10.1016/j.ijfoodmicro.2008.12.025
22. Perez-Rodriguez F, Valero A (2013) Predictive microbiology in foods. Springer, New York
23. Peñalver-Soto JL, Garre A, Esnoz A et al (2019) Guidelines for the design of (optimal) isothermal inactivation experiments. Food Res Int 126:108714. https://doi.org/10.1016/j.foodres.2019.108714
24. Poschet F, Vereecken KM, Geeraerd AH et al (2005) Analysis of a novel class of predictive microbial growth models and application to coculture growth. Int J Food Microbiol 100: 107–124. https://doi.org/10.1016/j.ijfoodmicro.2004.10.008
25. Aspridou Z, Koutsoumanis K (2020) Variability in microbial inactivation: from deterministic Bigelow model to probability distribution of single cell inactivation times. Food Res Int 137:109579. https://doi.org/10.1016/j.foodres.2020.109579
26. Peleg M, Cole MB (1998) Reinterpretation of microbial survival curves. Crit Rev Food Sci Nutr 38:353–380. https://doi.org/10.1080/10408699891274246
27. Mafart P, Couvert O, Gaillard S, Leguerinel I (2002) On calculating sterility in thermal preservation methods: application of the Weibull frequency distribution model. Int J Food Microbiol 72:107–113. https://doi.org/10.1016/S0168-1605(01)00624-9

28. Couvert O, Gaillard S, Savy N et al (2005) Survival curves of heated bacterial spores: effect of environmental factors on Weibull parameters. Int J Food Microbiol 101:73–81. https://doi.org/10.1016/j.ijfoodmicro.2004.10.048
29. Metselaar KI, den Besten HMW, Abee T et al (2013) Isolation and quantification of highly acid resistant variants of Listeria monocytogenes. Int J Food Microbiol 166:508–514. https://doi.org/10.1016/j.ijfoodmicro.2013.08.011
30. Geeraerd AH, Herremans CH, Van Impe JF (2000) Structural model requirements to describe microbial inactivation during a mild heat treatment. Int J Food Microbiol 59:185–209. https://doi.org/10.1016/S0168-1605(00)00362-7
31. Baranyi J, Roberts TA (1994) A dynamic approach to predicting bacterial growth in food. Int J Food Microbiol 23:277–294. https://doi.org/10.1016/0168-1605(94)90157-0
32. Hindmarsh A (1983) ODEPACK, a systematized collection of ODE solvers. In: Stepleman RS (ed) Imacs transactions on scientific computation. North-Holland, Amsterdam, p 372
33. Clemente-Carazo M, Cebrián G, Garre A, Palop A (2020) Variability in the heat resistance of Listeria monocytogenes under dynamic conditions can be more relevant than that evidenced by isothermal treatments. Food Res Int 137:109538. https://doi.org/10.1016/j.foodres.2020.109538
34. Corradini MG, Peleg M (2009) Dynamic model of heat inactivation kinetics for bacterial adaptation. Appl Environ Microbiol 75:2590–2597. https://doi.org/10.1128/AEM.02167-08
35. Dolan KD, Valdramidis VP, Mishra DK (2013) Parameter estimation for dynamic microbial inactivation: which model, which precision? Food Control 29:401–408. https://doi.org/10.1016/j.foodcont.2012.05.042
36. Garre A, González-Tejedor G, Peñalver-Soto JL et al (2018) Optimal characterization of thermal microbial inactivation simulating non-isothermal processes. Food Res Int 107:267–274. https://doi.org/10.1016/j.foodres.2018.02.040
37. Garre A, González-Tejedor GA, Aznar A et al (2019) Mathematical modelling of the stress resistance induced in Listeria monocytogenes during dynamic, mild heat treatments. Food Microbiol 84:103238. https://doi.org/10.1016/j.fm.2019.06.002
38. Hassani M, Cebrián G, Mañas P et al (2006) Induced thermotolerance under nonisothermal treatments of a heat sensitive and a resistant strain of Staphylococcus aureus in media of different pH. Lett Appl Microbiol 43:619–624. https://doi.org/10.1111/j.1472-765X.2006.02014.x
39. Stasiewicz MJ, Marks BP, Orta-Ramirez A, Smith DM (2008) Modeling the effect of prior sublethal thermal history on the thermal inactivation rate of salmonella in ground Turkey. J Food Prot 71:279–285. https://doi.org/10.4315/0362-028X-71.2.279
40. Garre A, Egea JA, Iguaz A et al (2018) Relevance of the induced stress resistance when identifying the critical microorganism for microbial risk assessment. Front Microbiol 9. https://doi.org/10.3389/fmicb.2018.01663
41. Huertas JP, Aznar A, Esnoz A et al (2016) High heating rates affect greatly the inactivation rate of Escherichia coli. Front Microbiol 7:1–8. https://doi.org/10.3389/fmicb.2016.01256
42. Huertas J-P, Ros-Chumillas M, Garre A et al (2021) Impact of heating rates on alicyclobacillus acidoterrestris heat resistance under non-isothermal treatments and use of mathematical modelling to optimize orange juice processing. Foods 10:1496. https://doi.org/10.3390/foods10071496
43. Janssen M, Verhulst A, Valdramidis V et al (2008) Inactivation model equations and their associated parameter values obtained under static acid stress conditions cannot be used directly for predicting inactivation under dynamic conditions. Int J Food Microbiol 128:136–145. https://doi.org/10.1016/j.ijfoodmicro.2008.06.009
44. Valdramidis VP, Geeraerd AH, Van Impe JF (2007) Stress-adaptive responses by heat under the microscope of predictive microbiology. J Appl Microbiol 103:1922–1930. https://doi.org/10.1111/j.1365-2672.2007.03426.x
45. Alvarenga VO, Gonzales-Barron U, do Prado Silva L et al (2021) Using extended Bigelow meta-regressions for modelling the effects of temperature, pH, Brix on the inactivation of heat resistant moulds. Int J Food Microbiol 338:108985. https://doi.org/10.1016/j.ijfoodmicro.2020.108985
46. Possas A, Pérez-Rodríguez F, Valero A, García-Gimeno RM (2017) Modelling the inactivation of Listeria monocytogenes by high hydrostatic pressure processing in foods: a review.

Trends Food Sci Technol 70:45–55. https://doi.org/10.1016/j.tifs.2017.10.006

47. Banga JR, Balsa-Canto E (2008) Parameter estimation and optimal experimental design. Essays Biochem 45:195–210. https://doi.org/10.1042/bse0450195
48. Garre A, Huertas JP, González-Tejedor GA et al (2018) Mathematical quantification of the induced stress resistance of microbial populations during non-isothermal stresses. Int J Food Microbiol 266:133–141. https://doi.org/10.1016/j.ijfoodmicro.2017.11.023
49. Garre A, Zwietering MH, den Besten HMW (2023) The importance of what we cannot observe: Experimental limitations as a source of bias for meta-regression models in predictive microbiology. International Journal of Food Microbiology 387:110045. https://doi.org/10.1016/j.ijfoodmicro.2022.110045
50. van Boekel MAJS (2021) To pool or not to pool: that is the question in microbial kinetics. Int J Food Microbiol 354:109283. https://doi.org/10.1016/j.ijfoodmicro.2021.109283
51. den Besten HMW, Wells-Bennik MHJ, Zwietering MH (2018) Natural diversity in heat resistance of bacteria and bacterial spores: impact on food safety and quality. Annu Rev Food Sci Technol 9:383–410. https://doi.org/10.1146/annurev-food-030117-012808
52. Jarvis B (2016) Statistical Aspects of the Microbiological Examination of Foods, 3rd ed. Academic Press. https://doi.org/10.1016/B978-0-12-803973-1
53. Garre A, Zwietering MH, den Besten HMW (2020) Multilevel modelling as a tool to include variability and uncertainty in quantitative microbiology and risk assessment. Thermal inactivation of listeria monocytogenes as proof of concept. Food Res Int 137:109374. https://doi.org/10.1016/j.foodres.2020.109374
54. Crespo Tapia N, Dorey AL, Gahan CGM et al (2020) Different carbon sources result in differential activation of sigma B and stress resistance in listeria monocytogenes. Int J Food Microbiol 320:108504. https://doi.org/10.1016/j.ijfoodmicro.2019.108504
55. Aryani DC, den Besten HMW, Zwietering MH (2016) Quantifying variability in growth and thermal inactivation kinetics of Lactobacillus plantarum. Appl Environ Microbiol 82:4896–4908. https://doi.org/10.1128/AEM.00277-16
56. Clemente-Carazo M, Leal J-J, Huertas J-P et al (2021) The different response to an acid shock of two salmonella strains marks their resistance to thermal treatments. Front Microbiol 12. https://doi.org/10.3389/fmicb.2021.691248
57. Verheyen D, Govaert M, Seow TK et al (2020) The complex effect of food matrix fat content on thermal inactivation of listeria monocytogenes: case study in emulsion and gelled emulsion model systems. Front Microbiol 10. https://doi.org/10.3389/fmicb.2019.03149
58. Possas A, Posada-Izquierdo GD, Zurera G, Pérez-Rodríguez F (2021) Evaluating the fate of Escherichia coli O157:H7 and *Salmonella* spp. on cucumbers. Food Microbiol 99:103830. https://doi.org/10.1016/j.fm.2021.103830
59. Akkermans S, Verheyen D, Smet C, Van Impe JFM (2021) A population balance model to describe the evolution of sublethal injury. Foods 10:1674. https://doi.org/10.3390/foods10071674
60. Conesa R, Andreu S, Fernández PS et al (2009) Nonisothermal heat resistance determinations with the thermoresistometer Mastia. J Appl Microbiology 107:506–513. https://doi.org/10.1111/j.1365-2672.2009.04236.x
61. Dolan KD, Mishra DK (2013) Parameter estimation in food science. Annu Rev Food Sci Technol 4:401–422. https://doi.org/10.1146/annurev-food-022811-101247
62. Fernández A, Ocio M, Fernández P et al (1999) Application of nonlinear regression analysis to the estimation of kinetic parameters for two enterotoxigenic strains ofBacillus cereus spores. Food Microbiol 16:607–613. https://doi.org/10.1006/fmic.1999.0282
63. Soetaert K, Petzoldt T, Setzer RW (2010) Solving differential equations in R: package deSolve. J Stat Softw 33. https://doi.org/10.18637/jss.v033.i09
64. Virtanen P, Gommers R, Oliphant TE et al (2020) SciPy 1.0: fundamental algorithms for scientific computing in python. Nat Methods 17:261–272. https://doi.org/10.1038/s41592-019-0686-2
65. Moré JJ (1978) The Levenberg-Marquardt algorithm: implementation and theory. In: Numerical analysis. Springer, Berlin/Heidelberg, pp 105–116
66. Haario H, Laine M, Mira A, Saksman E (2006) DRAM: efficient adaptive MCMC.

Stat Comput 16:339–354. https://doi.org/10.1007/s11222-006-9438-0

67. Soetaert K, Petzoldt T (2010) Inverse modelling, sensitivity and Monte Carlo analysis in R using package FME. J Stat Softw 33. https://doi.org/10.18637/jss.v033.i03
68. Balsa-Canto E, Henriques D, Gábor A, Banga JR (2016) AMIGO2, a toolbox for dynamic modeling, optimization and control in systems biology. Bioinformatics 32:3357–3359. https://doi.org/10.1093/bioinformatics/btw411
69. Bates DM, Watts D (2007) Nonlinear regression analysis and its applications, 1st edn. Wiley-Interscience, New York
70. Baty F, Ritz C, Charles S et al (2015) A toolbox for nonlinear regression in R : the package nlstools. J Stat Softw 66. https://doi.org/10.18637/jss.v066.i05
71. Draper NR, Smith H (1998) Applied regression analysis, 3rd rev edn. Wiley-Blackwell, New York
72. Aspridou Z, Koutsoumanis KP (2015) Individual cell heterogeneity as variability source in population dynamics of microbial inactivation. Food Microbiol 45:216–221. https://doi.org/10.1016/j.fm.2014.04.008
73. Garces-Vega F, Marks BP (2014) Use of simulation tools to illustrate the effect of data management practices for low and negative plate counts on the estimated parameters of microbial reduction models. J Food Prot 77:1372–1379. https://doi.org/10.4315/0362-028X.JFP-13-462
74. Garre A, Zwietering MH, van Boekel MAJS (2022) The Most Probable Curve method - A robust approach to estimate kinetic models from low plate count data resulting in reduced uncertainty. International Journal of Food Microbiology 380:109871. https://doi.org/10.1016/j.ijfoodmicro.2022.109871
75. Benford D, Halldorsson T, Jeger MJ et al (2018) Guidance on uncertainty analysis in scientific assessments. EFSA J 16. https://doi.org/10.2903/j.efsa.2018.5123
76. Nauta MJ (2000) Separation of uncertainty and variability in quantitative microbial risk assessment models. Int J Food Microbiol 57: 9–18. https://doi.org/10.1016/S0168-1605(00)00225-7
77. Vilas C, Arias-Méndez A, García MR et al (2016) Toward predictive food process models: a protocol for parameter estimation. Crit Rev Food Sci Nutr:1–14. https://doi.org/10.1080/10408398.2016.1186591
78. Ros-Chumillas M, Garre A, Maté J et al (2017) Nanoemulsified D-limonene reduces the heat resistance of Salmonella Senftenberg over 50 times. Nano 7:65. https://doi.org/10.3390/nano7030065
79. Vásquez GA, Busschaert P, Haberbeck LU et al (2014) An educationally inspired illustration of two-dimensional quantitative microbiological risk assessment (QMRA) and sensitivity analysis. Int J Food Microbiol 190:31–43. https://doi.org/10.1016/j.ijfoodmicro.2014.07.034
80. Huang L, Hwang C-A (2017) Dynamic analysis of growth of Salmonella Enteritidis in liquid egg whites. Food Control 80:125–130. https://doi.org/10.1016/j.foodcont.2017.04.044
81. Hwang C-A, Huang L (2019) Growth and survival of Bacillus cereus from spores in cooked rice – one-step dynamic analysis and predictive modeling. Food Control 96:403–409. https://doi.org/10.1016/j.foodcont.2018.09.036
82. Balsa-Canto E, Alonso-del-Real J, Querol A (2020) Temperature shapes ecological dynamics in mixed culture fermentations driven by two species of the Saccharomyces Genus. Front Bioeng Biotechnol 8. https://doi.org/10.3389/fbioe.2020.00915
83. Henriques D, Alonso-del-Real J, Querol A, Balsa-Canto E (2018) Saccharomyces cerevisiae and S. kudriavzevii synthetic wine fermentation performance dissected by predictive modeling. Front Microbiol 9. https://doi.org/10.3389/fmicb.2018.00088
84. Thornton A (2000) Publication bias in meta-analysis its causes and consequences. J Clin Epidemiol 53:207–216. https://doi.org/10.1016/S0895-4356(99)00161-4
85. Guillén S, Marcén M, Álvarez I et al (2020) Stress resistance of emerging poultry-associated Salmonella serovars. Int J Food Microbiol 335:108884. https://doi.org/10.1016/j.ijfoodmicro.2020.108884
86. Kooiman WJ, Barker AN, Gould GW, Wolf J (1973) The screw cap tube technique: a new and accurate technique for the determination of the wet heat resistance of bacterial spores. In: Spore Research. Academic Press, London, pp 87–92
87. Pflug IJ, Esselen W (1953) Development and application of an apparatus for study of thermal resistance of bacterial spores and thiamine at temperatures above 250-degrees-F. Food Technol 7:237 241

88. Brown KL, Gaze JE, McClement RH, Withers P (1988) Construction of a computer-controlled thermoresistometer for the determination of the heat resistance of bacterial spores over the temperature range 100 to 150°C. Int J Food Sci Technol 23:361–371. https://doi.org/10.1111/j.1365-2621.1988.tb00590.x
89. Condon S, Lopez P, Oria R, Sala FJ (1989) Thermal death determination: design and evaluation of a thermoresistometer. J Food Sci 54:451–457. https://doi.org/10.1111/j.1365-2621.1989.tb03104.x
90. Aryani DC, den Besten HMW, Hazeleger WC, Zwietering MH (2015) Quantifying strain variability in modeling growth of listeria monocytogenes. Int J Food Microbiol 208:19–29. https://doi.org/10.1016/j.ijfoodmicro.2015.05.006

Chapter 6

Growth/No-Growth Microbial Models in Food Science

Angie Dahiana Duque Rodriguez, Mírian Pereira da Silva, Natan de Jesus Pimentel-Filho, and Wilmer Edgard Luera Pena

Abstract

Logistic regression models quantify the occurrence of microbial development in a real scenario from production and commercialization to consumption of the product. Recent studies have demonstrated the versatility of its application as a tool to achieve microbiological food quality and safety. However, more knowledge of the methodology to be followed is needed, as well as the computer software available to be used. This chapter describes the mathematical principles, procedure, and validation to develop a growth/no-growth model and its possible applications in the various areas of food science.

Key words Probabilistic model, Logistic regression, Spoilage microorganisms, Pathogen microorganisms, Food quality, Mathematical model

1 Introduction

Food spoilage can be defined as a process or change that makes a food or food product undesirable or unacceptable for consumption. This complex phenomenon is the result of microbial activity and the occurrence of physicochemical processes in food constituents [1]. Likewise, safety is an important and essential component of quality and the first requirement of any food; therefore, any harmful microbial or chemical contaminants cannot be allowed in food consumption. Consequently, conservation techniques must prevent the growth of pathogenic microorganisms, as well as spoilage microorganisms [2]. In this context, predictive microbiology uses mathematical expressions to describe the behavior or response of microorganisms, that is, their multiplication, inhibition, or cell death, when they are submitted to intrinsic and/or intrinsic factors. As a result, microbial response parameters such as the specific growth rate, death rate, or the probability of growth are quantified. Thus, quantitative microbiology may be the most adequate

Verônica Ortiz Alvarenga (ed.), *Basic Protocols in Predictive Food Microbiology*, Methods and Protocols in Food Science,
https://doi.org/10.1007/978-1-0716-3413-4_6,

description of the approach for modeling the microbial ecology of food [3].

The conservation of most food products is achieved by using barriers, that is, several parameters such as water activity, pH, and storage temperature, among others, which together define the microbiological and physicochemical stability [4]. The main advantage of this technique is the synergistic effect, especially when the mechanisms and factors that govern microbial development in foods are known. This effect is possible when these barriers cause changes in cell structures and/or biomolecules, for example, in cell membranes, DNA, enzyme systems, pH, water activity (Aw), and electrical potential (Eh), among others, and therefore destabilizes the homeostasis of the microorganism, causing stress and later cell death [5]. Mathematical modeling combined with simulation and optimization offer several advantages, especially when developing processes and control strategies that significantly reduce the total time to solve problems and to reach the microbiological quality of the matrix and/or food during its different stages of production and its final production [6–8].

Among the models implemented in predictive microbiology, those that predict the probability of the development of a given microorganism when submitted to specific intrinsic and/or extrinsic factors stand out. When predictive microbiology started being developed, models currently known as probabilistic models made it possible to predict the growth risk of certain pathogenic microorganisms and/or risk of certain microbial metabolites, such as enzymes and toxins. Thus, the risk of the presence of microorganisms could be managed that were of interest to the food industry in certain food groups, thus ensuring that such microorganisms will not be present or will not be able to multiply in the final product [9].

The improvement of this type of model is based on logistic regression, a statistical technique that aims to produce (from statistical data), a model that can predict values taken by a categorical variable (a variable that can take on only a limited, and usually fixed, number of possible values, assigning each individual or other unit of observation to a given group or nominal category based on some qualitative property). It is often binary, from a series of continuous and/or binary explanatory variables, and thus determines the probability of an event occurring [10].

In general, probability models are applied to data that can be measured as "positive" or "negative." For example, if we consider the variable "toxin detection," only two answers are possible: "detectable" or "not detectable." The same can be said for the variable "growth detection." Later, this scenario led to the development of models based, for example, on the growth limits – no growth (C-NC) or *interface* modeling [11]. This type of modeling has been widely used by different researchers as an effective tool to determine the inhibition conditions of certain microorganisms [9, 10, 12–19] and thus allows the food industry to produce safe food.

2 Methods and Principles of the Logistic Regression Model

2.1 Developing Logistic Regression Model Equations

Regression methods have become an integral component of any data analysis concerned with describing the relationship between a response variable and one or more explanatory variables. The outcome variable is likely to be discrete, taking on two or more possible values. The logistic regression model is the most frequently used regression model to analyze these data [20]. Probabilistic modeling focuses on understanding microbial growth and no-growth, and probabilistic models are used to define the combinations of different conditions of factors involved in the process (barriers) that affect the probability of microbial growth [21–23].

Many distribution functions have been proposed for use when analyzing a dichotomous outcome variable [20]. Logistic regression modeling adopts data that can be measured as positive or negative. For example, when evaluating microbial growth, the response obtained can be coded as 1 (positive response) or 0 (negative response). Using regression techniques, the probability of occurrence (values between 0 and 1) of the studied phenomenon can be calculated. This probability can be related to intrinsic and extrinsic factors such as pH, Aw, and temperature [9].

Statistically, a logistic regression model relates the probability of occurrence of an event y, conditioned to a vector x of explanatory variables. Suppose there are n individual reports called events. These events can be the success or failure. Then, $yi = 1$ can be defined if the i-th event is a success and $yi = 0$ if it is a failure. Let us suppose now that for each of the n reports, p independent variables (x_{i1}, ...x_{ip}) can be measured, which can be qualitative or quantitative. The question is to consider how these independent variables influence the dependent variable y. The possible model would be the ordinary linear regression technique, which assumes that y_i are normally distributed with mean p_i and variance σ^2, where p_i is defined as the probability of success.

Numerically, the expression is as follows:

$$p_{i=p(y_i=1|x_{i1},\dots x_{ip})} \tag{1}$$

$$1-p_{i=p(y_i=0|x_{i1},\dots x_{ip}),i=1\dots n} \tag{2}$$

Equation 2 is linearly dependent of x_{ij} and the model can be described as follows:

$$p_i=\sum\nolimits_{j=1}^{p} b_j x_{ij} \tag{3}$$

The coefficients obtained indicate the relationships between the variables and the log odds in favor of success [24]. Consequently, a new value of p_i can be estimated by substituting the value of x_{ij} in Eq. 3. The probability of success in function of the independent variables can be given by Eqs. 4 and 5 [20]:

$$pi = \frac{\exp\left(\sum_{j=0}^{p} b_j x_{ij}\right)}{1 + \exp\left(\sum_{j=0}^{p} b_j x_{ij}\right)} \quad (4)$$

$$1 - pi = \frac{1}{1 + \exp\left(\sum_{j=0}^{p} b_j x_{ij}\right)} \quad (5)$$

where $x_i = (x_{i0} \ldots x_{ip})$, $x_{i0} = 1$, and b_j are unknown coefficients. The logarithm of the ratio of P_i and $1-Pi$ is a simple linear function of the $x_{ij's}$ and is called the logistic transform.

$$\ln\left(\frac{pi}{1 - pi}\right) = g(x) = \sum_{j=0}^{p} b_j x_{ij} \quad (6)$$

For a model with two independent variables (x_1 and x_2), the equation would be as follows:

$$\ln\left(\frac{pi}{1 - pi}\right) = g(x) = b_0 + b_1 x_1 + b_2 x_2 + b_3 x_1^2 + b_4 x_2^2 + b_5 x_1 x_2 \quad (7)$$

where $g(X)$ is called the logistic transformation (logit of p). The importance of this transformation is that $g(x)$ has many desirable properties of a linear regression model. The logit is linear in this parameter, it can be continuous, and it can vary depending on the range of x [20].

Another important difference between logistic and linear regression models refers to the conditional distribution of the response variable. In the linear regression model, it is assumed that the observation of the variable response can be expressed as $y = E(y/x) + \varepsilon$, where ε is called the error, which expresses the deviations of the observations around a conditional mean. A common assumption is that the error is normally distributed with a mean of zero, with some constant variation across the levels of the independent variable. Thus, a conditional distribution of the response variable can be normal with mean $E(y/x)$ and constant variance. This is not the case for a dichotomous variable. In this situation, the value of the response variable, given x, can be expressed as $y = p(x) + \varepsilon$, where ε can take in two possible values. If $y = 1$, then $\varepsilon = 1-p(X)$ with probability $p(x)$ and if $y = 0$, then $\varepsilon = -p(X)$ with probability $1-p(X)$. Thus, ε has a distribution with mean zero and variance equal to $p(x)(1-p(X))$ [20].

Many advantages are observed when using a logistic regression analysis approach; however, some problems were discussed by Ratkowsky, for example, convergence to a global optimum is often not achieved, or convergence may occur with an inappropriate set of conditions. However, the author concludes that the use of logistic regression modeling resulted in models with highly acceptable levels of agreement and good quality of fit for data on the presence/absence of microbial growth [25]. Other research on probabilistic models has shown the importance of the number of repetitions and concentrations necessary to correctly predict the behavior of the microorganism studied.

Thus, studies show that the use of a reduced number of replicates by combining environmental factors does not allow the convergence of an adequate dataset and, consequently, obtaining reliable conclusions [26].

2.2 Input Data for Logistic Regression Modeling

When using logistic regression modeling, the input data can result from different experimental designs. Lanciotti et al. studied the effects of pH, water activity (Aw), incubation temperature, and ethanol concentration on the growth/no-growth of *Bacillus cereus* FG1, *Salmonella enteritidis* B5, and *Staphylococcus aureus* S33 in brain heart infusion (BHI) broth and applied central composite designs (CCD) (*see* Chapter 1) for four factors and five levels (Table 1) [27]. The occurrence of growth was tested after 48 h of incubation by recording a visible increase in the turbidity of the broth or direct plating on specific growth media when the results were unreliable. Panagou et al. applied a full-factorial design for studied growth/no-growth interface of *Byssochlamys fulva* and *Byssochlamys nivea* on a synthetic medium as a function of temperature and water activity for a period of 30 days. The growth response was obtained from the colony diameter [28]. Recently Zalazar et al. used a full-factorial design with 96 treatments resulting from the combination of 6 levels of natamycin, 4 levels of NaCl, and 4 levels of xanthan gum obtaining a total of 976 observations, including replicates [19].

Several studies in the literature [10, 29–32] used this approach to obtain data in culture media simulating characteristics of food matrices. In these studies, different intrinsic and extrinsic parameters that affect the development of the microorganism were used, including the effect of the inoculum level on the growth/no-growth interface [32]. In this context, it can be believed that all factors that affect the growth/not growth interface affect the implemented model. These effects were pointed out by Ross and Dalgaard. They discussed the effect of temperature (during processing, storage, and food distribution); storage atmosphere (aerobically conditioned and controlled atmosphere packaging); salt; water-phase salt and water activity; pH; added preservatives including organic acids, nitrate, and spices; and smoke components [11].

Table 1
Experimental designs used to evaluate the growth/no-growth interfaces of *Salmonella enteritidis*, *Staphylococcus aureus*, and *Bacillus cereus*

	Experimental designs			
Run	A_w	Temperature (°C)	pH	Ethanol (%v/v)
1	0.915 (0.960)	17.5	4.75 (5.5)	1.5
2	0.915 (0.960)	17.5	6.25 (6.5)	0.5
3	0.915 (0.960)	32.5	4.75 (5.5)	0.5
4	0.915 (0.960)	32.5	6.25 (6.5)	1.5
5	0.965 (0.980)	17.5	4.75 (5.5)	0.5
6	0.965 (0.980)	17.5	6.25 (6.5)	1.5
7	0.965 (0.980)	32.5	4.75 (5.5)	1.5
8	0.965 (0.980)	32.5	6.25 (6.5)	0.5
9	0.940 (0.970)	25.0	5.50 (6.0)	1.0
10	0.940 (0.970)	25.0	5.50 (6.0)	1.0
11	0.915 (0.960)	17.5	4.75 (5.5)	0.5
12	0.915 (0.960)	17.5	6.25 (6.5)	1.5
13	0.915 (0.960)	32.5	4.75 (5.5)	1.5
14	0.915 (0.960)	32.5	6.25 (6.5)	0.5
15	0.965 (0.980)	17.5	4.75 (5.5)	1.5
16	0.965 (0.980)	17.5	6.25 (6.5)	0.5
17	0.965 (0.980)	32.5	4.75 (5.5)	0.5
18	0.965 (0.980)	32.5	6.25 (6.5)	1.5
19	0.940 (0.970)	25.0	5.50 (6.0)	1.0
20	0.940 (0.970)	25.0	5.50 (6.0)	1.0
21	0.890 (0.950)	25.0	5.50 (6.0)	1.0
22	0.990 (0.990)	25.0	5.50 (6.0)	1.0
23	0.940 (0.970)	10.0	5.50 (6.0)	1.0
24	0.940 (0.970)	40.0	5.50 (6.0)	1.0
25	0.940 (0.970)	25.0	4.00 (5.0)	1.0
26	0.940 (0.970)	25.0	7.00 (7.0)	1.0
27	0.940 (0.970)	25.0	5.50 (6.0)	0.0
28	0.940 (0.970)	25.0	5.50 (6.0)	2.0
29	0.940 (0.970)	25.0	5.50 (6.0)	1.0
30	0.940 (0.970)	25.0	5.50 (6.0)	1.0

Temperature and ethanol concentration were kept at the same levels in the two designs, while Aw and pH were changed. Levels in the second design are reported between parentheses. (Reproduced from Ref. [27] with permission from Elsevier)

Therefore, it is important to have a good experimental design in which the variables are well defined and the ranges of these variables meet the research objectives.

2.3 Adjustment in Statistical Software and Performance of Logistic Regression Models

The development of mathematical models in predictive microbiology goes through three levels: (1) primary models, which are used to describe the microbial growth process over time in a constant environment; (2) secondary models, which simulate the effects of environmental conditions on the kinetic parameters of the primary model; and (3) tertiary models, which combine primary and secondary models in a piece of software where it is possible to insert values of the input variables and obtain values of the dependent variable [33, 34].

The programs are widely used in predictive microbiology (Online Pathogen Modeling Program (Online PMP)) developed by the US Department of Agriculture and are available for free on the Internet (https://pmp.errc.ars.usda.gov/PMPOnline.aspx) and Combase [35] (https://www.combase.cc/index.php/en/). Unfortunately, they do not address growth/no-growth interface models. Recent research uses several programs to fit Eq. 1 to growth/no-growth data. The logistic regression modeling procedure of the Statistica software package (StatSoft, Inc., Tulsa, Okla, USA) was used in different research [27, 36]. The procedure regression logistic in SPSS Statistics (SPSS, Inc., Chicago, II, USA) was used by [12, 30, 31, 37]. The logistic procedure of SAS (Institute, Inc., Cary, NC, USA) was used by [13, 29, 32, 38, 39]. The R statistical software was used by different researchers [10, 19, 40]. The logistic regression procedure of Minitab (Minitab, Inc., State College, PA) was used by Morassi et al. and Panagou et al. [28, 41]. Moreover, the SYSTAT 12 software package (Systat Software, Inc., Washington, USA) was used by Arroyo-Lopez et al. in 2007 and 2012 [36, 42]. All these programs have a friendly graphical interface and are relatively easy to use.

Another procedure that has also been studied is neural networks. Describing the growth/no-growth of a bacterium requires a nonlinear approach; therefore, the neural network modeling technique can be used as an alternative method. Artificial neural networks (ANN) can be considered as a modeling technique inspired by biological nervous system architecture. This technique consists of many relatively simple systems that work in parallel to make decisions quickly. This system is structured in nodes connected through an artificial synapse symbolized by a matrix of numbers that can be adjusted by a learning process [43]. An application of this technique to fit growth/no-growth data was shown by Gunvig et al. [44]. The research aimed to develop a mathematical model to predict the growth/no-growth of psychrotrophic *Clostridium botulinum* in meat products packaged in a modified atmosphere with different combinations of pH, NaCl, sodium nitrite, and sodium

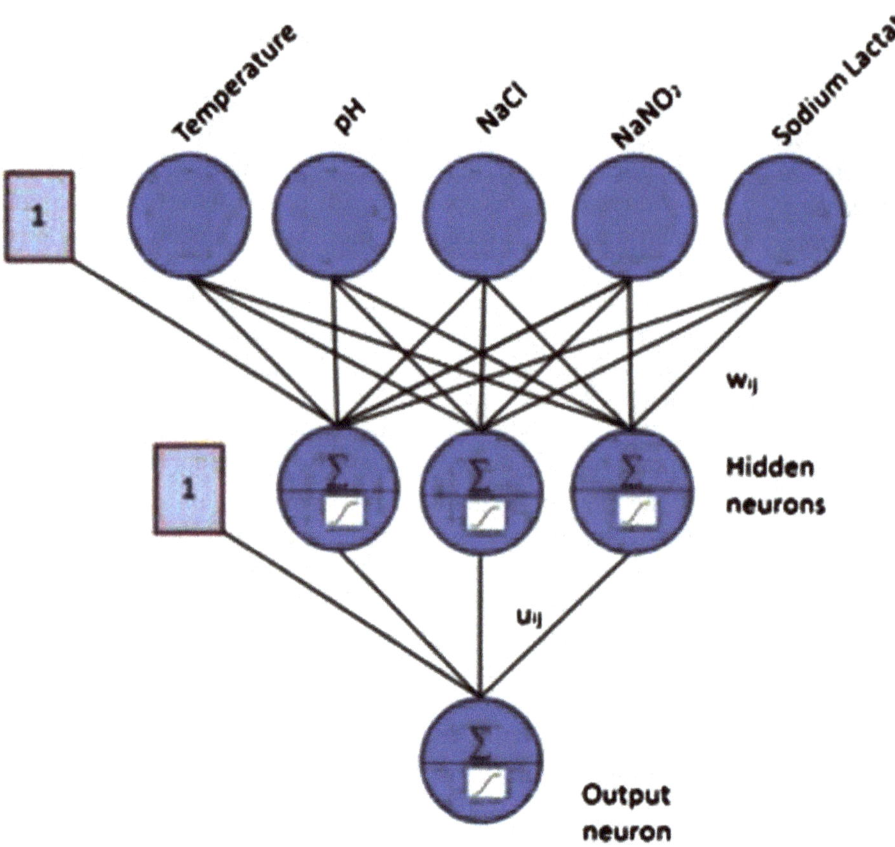

Fig. 1 The ANN architecture with three hidden neurons. (Reproduced from Ref. [44] with permission from Elsevier)

lactate stored at temperatures of 4 °C, 8 °C, and 12 °C for up to 8 weeks. An ANN was proposed with five input neurons (which correspond to the five variables: temperature, pH, NaCl, addition of sodium nitrite, and sodium lactate), three hidden, and one single output neuron (Fig. 1).

The ANN modeling process comprises two steps: (1) network training in which the largest amount of available data is used (e.g., 90%) and (2) validation in which the remaining data is used (e.g., 10%) [45]. The functioning of the ANN depends on mathematical functions associated with the nodes and layers, including sigmoidal function, hyperbolic tangent function, and Gaussian function for activation and the backpropagation algorithm for training [46, 47]. The performance seems to be favorable for ANN when compared to the Eq. 6 logistic model fitted in the statistical programs mentioned above. Other researchers also discussed the good performance of ANN to model microbial growth/no-growth data [48–51].

All logistic regression models developed must be tested using statistical validation indices. When developing growth/no-growth models, it must be verified whether they adequately explain the data. Indices such as Nagelkerke's R^2, Hosmer-Lemeshow adequacy test, and percentage agreement have been widely used for this purpose. The Hosmer-Lemeshow test used for binary response variables associates a chi-square value (Hosmer-Lemeshow chi-square) and a *p-value*; small *p*-values mean that the model is a poor fit. Thus, it can be observed that for a large *p-value*, the model fits the data well [20, 32]. In addition, the percentage of agreement and disagreement between observed and predicted values can be used. When the percentage of agreement is 100%, the predictive power of the model is perfect [9]. Nagelkerke's R^2 is an adjusted version of the Cox and Snell R^2 to achieve values from 0 to 1, as the Cox and Snell R^2 achieve values lower than 1 [52]. Another index used by Moraes et al. [53] was the receiver operating characteristic (ROC) curve that indicates the predictive capacity of the proposed logistic model, through sensitivity and specificity predictions, visualizing, organizing, and classifying the model based on its predictive performance [54]. All these indices are calculated by the software used to adjust logistic regression models. To validate the implemented models, some studies, such as [12, 41, 53], involved experiments in conditions different from those used to construct the logistic regression model to test its predictive power.

2.4 Stepwise for the Development the Growth/No-Growth Model

1. Data (*see* Subheading 2.2). The data matrix will be formed by intrinsic and/or extrinsic parameters (independent variables) according to the experimental designs and the response of the microorganism observed experimentally (growth (1) or no-growth (0)) (dependent variable). This data matrix can be written in an Excel spreadsheet.
2. Model fit: As an example, we will use the SPSS v12 program.
3. Open the SPSS for Windows application.
4. Click on "File" and then on "New" and finally "Data" (generates a date spreadsheet editable).
5. Copy data from Excel and paste it into the date spreadsheet.
6. Name the studied variables by clicking on the top of each data column.
7. Click on "Statistics" after "Regression" and finally "Binary Logistic" (generates the "Logistic Regression" window).
8. Select the dependent variable in the field "Dependent" and the independent variables in the field "Covariates" (here, you can put the independent variables in linear form, quadratically, and their interactions).

9. In "Options", to mark: Classifications plots, Hosmer-Lomeshow goodness-of-fit, include constant in model, Classification cutoff: 0.5, Maximum Iterations 20, CI for exp.(B): 95% (confidence interval of each estimated coefficient) and Method: Enter and finally Click on Ok.
10. The Logistic Regression window is generated. In this window, it is possible to observe all the model adjustment information, such as coefficient values of each variable (and standard deviation, p-value) and statistical indices of the adjusted model as R2-Negelkerke, R2-Cox and Snell, −2 Log Likelihood, and Hosmer-Lemeshow goodness-of-fit test. Independent variables that do not significantly influence the dependent variable must be excluded of the model, with the exception of linear variables in which their interactions are significant.
11. The procedure using other programs as Statistica and Minitab is similar.
12. With these results, it is possible to obtain the logistic regression equation (Eqs. 4 and 7) and predict probabilities for the dependent variable as well as critical values of the independent variables for a given probability value. Using graphics program, a response surface figure can also be generated.

3 Logistic Regression Modeling Application

Due to the advancement of predictive microbiology, over the last 20 years, different approaches using logistic regression models, such as modeling the interface between "growth" and "nonmicrobial growth," "biofilm formation" and "non-biofilm formation," and "toxin production" and "non-toxin production," have been proposed [12, 55, 56].

One of the first studies reporting the use of logistic regression models in the literature was by Salter et al. where temperature and salt (NaCl) limits were combined to predict the growth of a pathogenic *Escherichia coli* strain [57]. In the same year, Tienungoon et al. evaluated the growth limits of Listeria monocytogenes as a function of temperature, pH, NaCl, and lactic acid [58]. The same methodology was applied by López-Malo and Palou, to model the growth/no-growth of Zygosaccharomyces bailii in mango puree [59]. Until then, models had been limited to microbial growth/no-growth analyses. Considering the rapid development of logistic regression models, and due to recent interest, new approaches have emerged, such as the probability assessment of the adhesion and biofilm formation, enterotoxin production, compound formation, and product deterioration, among others.

Logistic regression has been widely used to estimate the shelf life of food products and for hazard analysis, identification and establishment of critical control points, and product research and development, among other applications of great importance in food science [60].

3.1 Microbial Shelf Life Studies

Predicting the shelf life of a food product is advantageous to minimize the risk of losses, extend shelf life, and maintain quality during food processing and distribution. Studies have been carried out on a variety of food products to determine their shelf life [61]. In order to determine the shelf life and microbiological stability of soft drinks, Belletti et al. evaluated the antimicrobial activity of aromatic compounds against Saccharomyces cerevisiae by logistic regression [62]. The deterioration probability of the beverage was evaluated as a function of the duration of heat treatment and concentration of flavorings and yeast inoculum, and the models obtained proved to be accurate for the three substances studied. In another approach, Koseki et al. studied the recovery of L. monocytogenes on sliced cooked ham after processing it at high pressure ranging from 400 to 600 MPa, pressure-holding time (1, 5, 10, 20, 30, 45, and 60 min), and the inoculum level (3 and 5 log 10 CFU/g), which provided adequate processing conditions and estimated the risk of L. monocytogenes recovery during storage [63]. The predicted interface of recovery/no recovery reflected the experimental observations. Similarly, Valdramidis et al. evaluated the recovery of L. monocytogenes in meats cured under high pressure (450–800 MPa) and sodium chloride (0–34.9 g/L) and sodium nitrite (mg/L) concentrations, where the recovery of L. monocytogenes was influenced by the applied pressure, the storage time, and the synergistic effect of pressure and *Aw* on inhibiting microbial recovery [64].

In an attempt to develop predictive models of growth/no-growth of spoilage molds on bread subjected to treatments with acetic acid and phenyllactic acid, Debonne et al. demonstrated the predictive potential use of the logistic regression model to predict the shelf life of par-baked bread, in which acetic acid prevented mold growth for 45 days [65].

3.2 Hazard Analysis and Critical Control Points (HACCP)

The implementation of a HACCP system depends on identifying critical control points of a given hazard to prevent potential food safety problems during food production, storage, and distribution [66]. Predictive models describe the influence of factors such as temperature, pH, salt concentration, and Aw on microbial behavior [67]. In addition, from logistic regression models, critical limits of factors related to microbial behavior can be established, which poses a risk to food safety. Thus, these models can be used to assist in the preliminary analysis of hazards, identifying and establishing critical control points and corrective actions to be taken [68].

To predict the limits of growth and ochratoxin production by Aspergillus carbonarius in pistachio nuts, Marín et al. used the logistic regression model. Pistachios were stored at different moisture contents (2.5%, 17.9%, 24.0%, 29.5%, and 34.8%) and temperatures (10, 15, 20, 25, 30, 37, and 42 °C). Pistachios have been shown to be kept below 10% moisture for up to 20 °C and at a moisture content as high as 25% under very cool conditions (<10 °C). The model was highly compatible with the experimental data and can thus be used to implement HACCP plans for the ochratoxins in pistachio prevention [69].

The adhesion/biofilm formation boundaries of *Salmonella enterica* as a function of pH (4, 5, 6, and 7), temperature (8 °C, 12 °C, 20 °C, and 35 °C), and NaCl concentration (0%, 2%, 4%, 6%, 8%, 10% w/v) were predicted by means of the probabilistic model using logistic regression. The ability for adhesion or biofilm formation varied among the *S. enterica* strains tested. It was reduced at temperatures <12 °C and increased at temperatures around 35 °C, decreased with the increase of NaCl concentration, and increased when pH increased. The high compatibility of the values predicted by the models developed with the values in assays of the experimental validation showed the great predictive power of the obtained models [53].

During the manufacture of meat products, *S. aureus* growth is a critical control point in HACCP plans. Thus, Gunvig et al. developed a mathematical model of logistic regression to predict the increase in the number of S. aureus and the formation of staphylococcal enterotoxin during heating and fermentation of meat products. The experiments involved different temperatures (10–40 °C), pH (4.6–6.0), salt (2.2–5.6%), and sodium nitrite concentrations (0–150 ppm). The salt and nutrient concentration factors did not influence the enterotoxin production. The formation of enterotoxins was more significant at temperatures above 30 °C and detected between 8 and 30 h only in the range of five to seven logs of S. aureus/g. The models showed agreement with the experimental data and are thus applicable to predict the increase in S. aureus and to evaluate the possibility of the staphylococcal enterotoxin formation while processing meat products [70].

Proper hazard analysis and critical control points are relevant for the food industry to achieve high quality and ensure food safety [68]. Therefore, the potential of developing predictive logistic regression models to help create HACCP systems is a topic of research worldwide [71].

3.3 Research and Development of New Products

The development of new products, formulations, or reformulations may affect their microbiological characteristics. Using growth/survival models to develop food products is advantageous, as the consequences of microbial behavior arising from formulations and changes in compositions or processing can be evaluated quickly and efficiently [72].

Predictive microbial models using logistic regression to ensure microbiological safety in new or reformulated products have been reported. Probability growth models of 5 enterotoxigenic strains of *Staphylococcus aureus* as a function of temperature (8, 10, 13, 16, and 19 °C), pH (4.5, 5.0, 5.5, 6.0, 6.5, 7.0, and 7.5), and Aw (19 levels ranging from 0.867 to 0.999) were developed. Different combinations of temperature, pH, and *Aw* largely influenced the growth of *S. aureus* boundaries. Growth of *S. aureus* was inhibited at 8 °C by appropriately lowering the pH and *Aw* levels. *S. aureus* resisted to low Aw values (0.867) and grew at low pH (4.5) [73]. The logistic regression model provided an adjustment of the observed data and can be a powerful tool to decide the most suitable ranges of environmental factors to design safe formulations for foods susceptible to *S. aureus* contamination.

Huang et al. proposed a logistic model to define the growth and no-growth boundary of *L. monocytogenes* with sodium tripolyphosphate (STPP), sodium lactate (NaL), sodium diacetate (NaDiAc), sodium chloride (NaCl), sodium nitrite ($NaNO_2$), and pH as control factors. The following levels of treatments were used: STPP (0–1500 ppm), NaL (0–4.8%), NaCl (0–8%), NaDiAc (0–2500 ppm), $NaNO_2$ (0–1800 ppm), and pH (4.5–8.5). The maximum of NO_2 (1.800 ppm) and NaDiAc (2.500 ppm) were found not effective in preventing the growth when applied alone. STPP was shown to be effective in preventing the growth of *L. monocytogenes.* It grew at pH 6–7 and was inhibited only at high concentrations of NaL and NaCl [29]. It was concluded that the logistic regression model developed in the study can be used to examine the effect of studied factors on growth of *L. monocytogenes* and to guide the development of proper formulation to prevent its growth in ready-to-eat food products and to reduce the risk of foodborne listeriosis.

Finally, probability models to control the spoilage of chocolate cake by *Penicillium citrinum* LMQA_053 at Aw values (0.78, 0.80, 0.83, and 0.86), pH (6.0, 6.3, 6.6, and 7.0), calcium propionate (0%, 0.10%, 0.15%, and 0.20%), and potassium sorbate (0%, 0.05%, and 0.10%) were developed [41]. To prevent the growth of *P. citrinum* LMQA-053, a low a_w (<0.78) and the highest concentration of preservatives (0.1% potassium sorbate, 0.2% calcium propionate) studied can be used to formulate chocolate cake in all pH and temperatures for 45 days of storage. The probabilistic model presents a good fit to the experimental data and can be used to develop or change formulas toward avoiding mold growth and spoilage of cakes.

References

1. Nychas GE, Panagou E (2011) Microbiological spoilage of foods and beverages. In: Food and beverage stability and shelf-life, Woodhead Publishing series in food science, technology and nutrition. Woodhead Publishing, pp 3–28. https://doi.org/10.1533/9780857092540.1.3
2. Alzamora SM, López-Malo A (2002) Microbial behavior modeling as a tool in the design and control of minimally processed foods. In: Engineering and food for the 21st century. CRC Press, Boca Raton, pp 661–680
3. Ross T, McMeekin TA (1995) Predictive microbiology and HACCP. In: HACCP in meat, poultry, and fish processing. Springer, Boston, pp 330–357
4. Rahman MS (2015) Hurdle technology in food preservation. In: Minimally processed foods. Springer, Cham, pp 17–33
5. Leistner L (2000) Basic aspects of food preservation by hurdle technology. Int J Food Microbiol 55(1–3):181–186. https://doi.org/10.1016/S0168-1605(00)00161-6
6. Garcia D, Ramos AJ, Sanchis V, Marín S (2009) Predicting mycotoxins in foods: a review. Food Microbiol 26(8):757–769. https://doi.org/10.1016/j.fm.2009.05.014
7. Prandini A, Sigolo S, Filippi L, Battilani P, Piva G (2009) Review of predictive models for Fusarium head blight and related mycotoxin contamination in wheat. Food Chem Toxicol 47(5):927–931. https://doi.org/10.1016/j.fct.2008.06.010
8. Zwietering MH, den Besten HM (2011) Modelling: one word for many activities and uses. Food Microbiol 28(4):818–822. https://doi.org/10.1016/j.fm.2010.04.015
9. Carrasco E, del Rosal S, Racero JC, García-Gimeno RM (2012) A review on growth/no growth *Salmonella* models. Food Res Int 47(1):90–99. https://doi.org/10.1016/j.foodres.2012.01.006
10. Tsuruma N, Doto S, Ishida W, Koyama K, Koseki S (2021) How many repetitions per condition are required for developing a stable growth/no growth boundary model for *Bacillus simplex* spores? Food Control 122:107756. https://doi.org/10.1016/j.foodcont.2020.107756
11. Ross T, Dalgaard P (2004) Secondary models. In: McKellar RB, Lu X (eds) Modelling microbial responses in food. CRC Press, Boca Raton, pp 63–150
12. Deschuyffeleer N, Vermeulen A, Daelman J, Castelein E, Eeckhout M, Devlieghere F (2015) Modelling of the growth/no growth interface of *Wallemia sebi* and *Eurotium herbariorum* as a function of pH, aw and ethanol concentration. Int J Food Microbiol 192:77–85. https://doi.org/10.1016/j.ijfoodmicro.2014.09.022
13. Huang L, Li C, Hwang CA (2018) Growth/no growth boundary of *Clostridium perfringens* from spores in cooked meat: a logistic analysis. Int J Food Microbiol 266:257–266. https://doi.org/10.1016/j.ijfoodmicro.2017.12.010
14. Gradeci K, Labonnote N, Time B, Köhler J (2017) A probabilistic-based approach for predicting mould growth in timber building envelopes: comparison of three mould models. Energy Procedia 132:393–398. https://doi.org/10.1016/j.egypro.2017.09.641
15. Masana MO, Baranyi J (2000) Growth/no growth interface of *Brochothrix thermosphacta* as a function of pH and water activity. Food Microbiol 17:485–493. https://doi.org/10.1006/fmic.2000.0349
16. McMeekin TA, Presser K, Ratkowsky D, Ross T, Salter M, Tienungoon S (2000) Quantifying the hurdle concept by modeling the bacterial growth/no growth interface. Int J Food Microbiol 55:93–98. https://doi.org/10.1016/S0168-1605(00)00182-3
17. Polese P, Del Torre M, Spaziani M, Stecchini ML (2011) A simplified approach for modelling the bacterial growth/no growth boundary. Food Microbiol 28(3):384–391. https://doi.org/10.1016/j.fm.2010.09.011
18. Schaffner DW, Labuza TP (1997) Predictive microbiology: where are we, and where are we going? Food Technol 51:95–99
19. Zalazar AL, Gliemmo MF, Soria M, Campos CA (2019) Modelling growth/no growth interface of *Zygosaccharomyces bailii* in simulated acid sauces as a function of natamycin, xanthan gum and sodium chloride concentrations. Food Res Int 116:916–924. https://doi.org/10.1016/j.foodres.2018.09.028
20. Hosmer DW, Emeshow S, Sturdivant RX (2013) Applied logistic regression, 3rd edn. Wiley, Hoboken, p 518
21. Lopez-Malo A, Palou E (2000) Modeling the growth/no-growth interface of *Zigosaccharomyces bailii* in mango puree. J Food Sci 65(3):516–520. https://doi.org/10.1111/j.1365-2621.2000.tb16039.x
22. Lopez-Malo A, Guerrero S, Alzamora SM (2000) Probabilistic modeling of *Saccharomyces cerevisiae* inhibition under the effects of

water activity, ph and potassium sorbate concentration. J Food Prot 63(1):91–95. https://doi.org/10.4315/0362-028X-63.1.91

23. Vermeulen A, Daelman J, Steenkiste JV, Devlieghere F (2012) Screening of different stress factors and development of growth/no growth models for *Zigosaccharomyces rouxii* in modified Saboraund medium, mimicking intermediate moisture foods (IMF). Food Microbiol 32: 389–396. https://doi.org/10.1016/j.fm.2012.07.019
24. Lee ET, Wang JW (2003) Statistical methods for survival data analysis, 3rd edn. Wiley, Hoboken
25. Ratkowsky DA (2002) Some examples of, and some problems with, the use of nonlinear logistic regression in predictive microbiology. Int J Food Microbiol 73:119–125. https://doi.org/10.1016/S0168-1605(01)00643-2
26. Valero M, Hernández-Herrero LA, Giner MJ (2007) Survival, isolation and characterization of a psychrotrophic *Bacillus cereus* strain from a mayonnaise-based ready-to-eat vegetable salad. Food Microbiol 24:671–677. https://doi.org/10.1016/j.fm.2007.04.005
27. Lanciotti R, Sinigaglia M, Gardini F, Vannini L, Guerzoni ME (2001) Growth/no growth interfaces of *Bacillus cereus*, *Staphylococcus aureus* and *Salmonella enteritidis* in model systems based on water activity, pH, temperature and etanol concentration. Food Microbiol 18: 659–668. https://doi.org/10.1006/fmic.2001.0429
28. Panagou EZ, Chelonas S, Chatzipavlidis I, Nychas GE (2010) Modeling the effect of temperature and water activity on the growth rate and growth/no growth interface of *Bysschlamys fulva* and *Bysschlamys nivea*. Food Microbiol 27:618–627. https://doi.org/10.1016/j.fm.2010.02.005
29. Huang L, Jia Z, Hwang CA (2022) Growth and no-growth boundary of *Listeria monocytogenes* in beef – a logistic modeling. Food Res Int 152:110919. https://doi.org/10.1016/j.foodres.2021.110919
30. Sun R, Vermeulen A, Devlieghere (2021) Modeling the combined effect of temperature, pH, acetic and lactic acid concentrations on the growth/no growth interface of acid-tolerant *Bacillus* spores. Int J Food Microbiol 360: 109419. https://doi.org/10.1016/j.ijfoodmicro.2021.109419
31. Debonne E, Vermeulen A, Bockstaele FV, Soljic I, Eeckhout M, Devlieghere F (2019) Growth/no-growth models of *in-vitro* growth of *Penicillium paneum* as a function of thyme essential oil, pH, aw, temperature. Food Microbiol 83:9–17. https://doi.org/10.1016/j.fm.2019.04.003
32. Vermeulen A, Gysemans KPM, Bernaerts K, Geeraerd AH, Dbevere J, Devlieghere F, Van Impe JF (2009) Modelling the influence of the inoculation level on the growth/no growth interface of *Listeria monocytogenes* as a function of pH, aw and acetic acid. Int J Food Microbiol 135:83–89. https://doi.org/10.1016/j.ijfoodmicro.2009.07.038
33. Huang L (2017) Dynamic identification of growth and survival kinetic parameters of microorganisms in foods. Curr Opin Food Sci 14:85–92. https://doi.org/10.1016/j.cofs.2017.01.013
34. Yue X, Sui J, Niu T, Liu Y, Liu X (2011) Modeling the effect of temperature and water activity on the growth rate and lag phase of *Aspergillus flavus* during rice drying. Dry Technol 29:1306–1312. https://doi.org/10.1080/07373937.2011.592039
35. Baranyi J, Tampli ML (2004) ComBase: a common database on microbial response to food environments. J Food Prot 67:1967–1971. https://doi.org/10.4315/0362-028X-67.9.1967
36. Arroyo-López FN, Buatista-Gallego J, Romero-Gil V, Rodriguez-Gomez F, Garrido-Fernandez A (2012) Growth/no growth interfaces of table olive related yeasts for natamycin, citric acid and sodium chloride. Int J Food Microbiol 155:257–262. https://doi.org/10.1016/j.ijfoodmicro.2012.02.007
37. Debonne E, Maene P, Vermeulen A, Bockstaele FV, Depredomme L, Vermeir P, Eeckhout M, Devlieghere F (2020) Validation of in-vitro antifungal activity of the fermentation quotient on bread spoilage moulds through growth/no growth modelling and bread baking trials. LWT-Food Sci Technol. https://doi.org/10.1016/j.lwt.2019.108636
38. Dang TDT, Mertens L, Vermeulen A, Geeraerd AH, Van Impe JF, Debevere J, Dvlieghere F (2010) Modelling the growth/no growth boundary of *Zigosaccharomyces bailii* in acidic conditions: a contribution to the alternative method to preserve foods without using chemical preservatives. Int J Food Microbiol 137:1–12. https://doi.org/10.1016/j.ijfoodmicro.2009.10.020
39. Skandamis PN, Stoforth JD, Kendall PA, Belk KE, Scanga JA, Smith GC, Sofos JN (2007) Modeling the effect of inoculum size and acid adaptation on growth/no growth interface of *Escherichia coli* O157: H7. Int J Food Microbiol 120:237–249. https://doi.org/10.1016/j.ijfoodmicro.2007.08.028

40. Haberbeck LU, Oliveira RC, Vivijs B, Wenseleers T, Aertsen A, Michiels C, Geeraerd AH (2015) Variability in growth/no growth boundaries of 188 different *Escherichia coli* strains reveals that approximately 75% have a higher growth probability under low pH conditions that *E. coli* O157:H7 strain ATCC 43888. Food Microbiol 45:222–230. https://doi.org/10.1016/j.fm.2014.06.024
41. Morassi LLP, Silva BS, Furtado MM, Freire L, Santos JLP, Chaves RD, Granato D, Silva MP, Peña WEL, Sant'Ana AS (2022) Growth/no growth modeling to control the spoilage of chocolate cake by *Penicillum citrinum* LMQA_053: impact of pH, water activity, temperature, and different concentration of calcium propionate and potassium sorbate. Food Control. https://doi.org/10.1016/j.foodcont.2022.109064
42. Arroyo-López FNA, Quintana MCD, Fernandez AG (2007) Modelling of the growth-no growth interface of *Issatchenkia occidentalis* an olive spoilage yeast, as a function of the culture media, NaCl, citric and sorbic acid concentrations: study of its inactivation in the no growth region. Int J Food Microbiol 117:150–159. https://doi.org/10.1016/j.ijfoodmicro.2007.03.005
43. Sreekanth S, Ramaswamy HS, Sablani S (2007) Prediction of Psychrometric parameters using neural networks. Dry Technol 16:825–837. https://doi.org/10.1080/07373939808917438
44. Gunvig A, Hansen F, Borggaard C (2013) A mathematical model for predicting growth/no growth of psychrotrophic *C. botulinum* in meat products with five variables. Food Control 29:309–317. https://doi.org/10.1016/j.foodcont.2012.06.046
45. Yu C, Davison VJ, Yang SX (2006) A neural network approach to predict survival/death and growth/no-growth interfaces for *Escherichia coli* O157:H7. Food Microbiol 23:552–560. https://doi.org/10.1016/j.fm.2005.09.008
46. Baughman DR, Liu YA (1995) Neural networks in bioprocessing and chemical engineering. Academic Press, San Diego, p 488
47. Demuth H, Beale M (2003) Neural network toolbox. The MathWorks, Natic, MA, p 509
48. Fernández JC, Hervás C, Martínez-Estudillo FJ, Gutierrez PA (2011) Memetic pareto evolitionary artificial neural networks to determine growth/no grotwh in predictive microbiology. Appl Soft Comput 11:534–550. https://doi.org/10.1016/j.asoc.2009.12.013
49. Fernandez-Navarro F, Valero A, Hervás-Martínez C, Gutierrez PA, Garcia-Gimeno RM, Zurera-Cosano G (2010) Deveplopment of a multi-classification neural network model to determine the microbial growth/no growth interface. Int J Food Microbiol 141:203–212. https://doi.org/10.1016/j.ijfoodmicro.2010.05.013
50. Hajmeer M, Basheer I (2002) A probabilistic neural network approach for modeling and classification of bacterial growth/no growth data. J Microbiol Methods 51:217–226. https://doi.org/10.1016/S0167-7012(02)00080-5
51. Hajmeer MN, Basheer IA (2003) A hybrid Bayesian-neural network approach for probabilistic modeling of bacterial growth/no growth interface. Int J Food Microbiol 82:233–243. https://doi.org/10.1016/S0168-1605(02)00308-2
52. Bewick V, Cheek L, Ball J (2005) Statistic review 14: logistic regression. Crit Care 9:112–118
53. Moraes JO, Cruz EA, Souza EGF, Oliveira TCM, Alvarenga VO, Peña WEL, Sant'Ana AS, Magnani M (2018) Prediction adhesion and biofilm formation boundaries on stainless steel surfaces by five *Salmonella enterica* strains belonging to different serovars as a function of pH, temperature and NaCl concentration. Int J Food Microbiol 281:90–100. https://doi.org/10.1016/j.ijfoodmicro.2018.05.011
54. Fawcett T (2006) An introduction to ROC analysis. Pattern Recogn Lett 27:861–874
55. Elahi S, Fujikawa H (2019) Comprehensive study of the boundaries of enterotoxin A production and growth of Staphylococcus aureus at various temperatures and salt concentrations. J Food Sci 84:121–126. https://doi.org/10.1111/1750-3841.14402
56. Moraes JO, Cruz EA, Pinheiro Í et al (2019) An ordinal logistic regression approach to predict the variability on biofilm formation stages by five *Salmonella enterica* strains on polypropylene and glass surfaces as affected by pH, temperature and NaCl. Food Microbiol 83:95–103. https://doi.org/10.1016/j.fm.2019.04.012
57. Salter MA, Ratkowsky DA, Ross T, McMeekin TA (2000) Modelling the combined temperature and salt (NaCl) limits for growth of a pathogenic *Escherichia coli* strain using nonlinear logistic regression. Int J Food Microbiol 61:159–167. https://doi.org/10.1016/S0168-1605(00)00352-4
58. Tienungoon S, Ratkowsky DA, McMeekin TA, Ross T (2000) Growth limits of *Listeria monocytogenes* as a function of temperature, pH, NaCl, and lactic acid. Appl Environ Microbiol

66:4979–4987. https://doi.org/10.1128/AEM.66.11.4979-4987.2000

59. López-Malo A, Palou E (2000) Modeling the growth/no-growth interface of *Zygosaccharomyces bailii* in mango puree. J Food Sci 65: 516–520. https://doi.org/10.1111/j.1365-2621.2000.tb16039.x
60. Messens W, Hempen M, Koutsoumanis K (2018) Use of predictive modelling in recent work of the panel on biological hazards of the European food safety authority. Microb Risk Anal 10:37–43. https://doi.org/10.1016/j.mran.2018.04.002
61. Taormina PJ, Taormina PJ (2021) Purposes and principles of shelf life determination. In: Taormina PJ, Hardin MD (eds) Food safety and quality-based shelf life of perishable foods. Springer, Cham, pp 1–26
62. Belletti N, Kamdem SS, Patrignani F et al (2007) Antimicrobial activity of aroma compounds against Saccharomyces cerevisiae and improvement of microbiological stability of soft drinks as assessed by logistic regression. Appl Environ Microbiol 73:5580–5586. https://doi.org/10.1128/AEM.00351-07
63. Koseki S, Mizuno Y, Yamamoto K (2007) Predictive modelling of the recovery of *Listeria* monocytogenes on sliced cooked ham after high pressure processing. Int J Food Microbiol 119:300–307. https://doi.org/10.1016/j.ijfoodmicro.2007.08.025
64. Valdramidis VP, Patterson MF, Linton M (2015) Modelling the recovery of *Listeria* monocytogenes in high pressure processed simulated cured meat. Food Control 47:353–358. https://doi.org/10.1016/j.foodcont.2014.07.022
65. Debonne E, Vermeulen A, Bouboutiefski N et al (2020) Modelling and validation of the antifungal activity of DL-3-phenyllactic acid and acetic acid on bread spoilage moulds. Food Microbiol 88:103407. https://doi.org/10.1016/j.fm.2019.103407
66. Stavropoulou E, Bezirtzoglou E (2019) Predictive Modeling of microbial behavior in food. Foods 8(12):654. https://doi.org/10.3390/foods8120654
67. Tarlak F, Ozdemir M, Melikoglu M (2018) Mathematical modelling of temperature effect on growth kinetics of *Pseudomonas* spp. on sliced mushroom (*Agaricus bisporus*). Int J Food Microbiol 266:274–281. https://doi.org/10.1016/j.ijfoodmicro.2017.12.017
68. McDonald K, Sun DW (1999) Predictive food microbiology for the meat industry: a review. Int J Food Microbiol 52:1–27. https://doi.org/10.1016/S0168-1605(99)00126-9
69. Marín S, Hodžić I, Ramos AJ, Sanchis V (2008) Predicting the growth/no-growth boundary and ochratoxin A production by *Aspergillus carbonarius* in pistachio nuts. Food Microbiol 25:683–689. https://doi.org/10.1016/j.fm.2008.03.006
70. Gunvig A, Andresen MS, Jacobsen T, Borggaard C (2018) Staphtox predictor – a dynamic mathematical model to predict formation of Staphylococcus enterotoxin during heating and fermentation of meat products. Int J Food Microbiol 285:81–91
71. Fakruddin M, Mazumder RM, Shahnewaj K, Mannan B (2011) Predictive microbiology: modeling microbial responses in food. Ceylon J Sci (Bio Sci) 40:121–131
72. Nakashima S, André C, Franco B (2000) Revisão: aspectos básicos da Microbiologia. Braz J Food Technol 3:41–51
73. Valero A, Pérez-Rodríguez F, Carrasco E et al (2009) Modelling the growth boundaries of Staphylococcus aureus. Effect of temperature, pH and water activity. Int J Food Microbiol 133:186–194. https://doi.org/10.1016/j.ijfoodmicro.2009.05.023

Chapter 7

Individual Cell-Based Modeling for Microbial Growth and Inactivation Using Time-Lapse Microscopy

Zafeiro Aspridou, Alexandra Lianou, and Konstantinos P. Koutsoumanis

Abstract

Most microbial growth, survival, and inactivation studies rely on large microbial populations, with the behavioral heterogeneity of individual cells not being considered. Given the importance of variability in risk-based food safety and quality management, the role of individual cell heterogeneity as a source of variability in predictive models has been recognized from many researchers. The methodology described herein allows for the observation of the behavior of individual microbial cells and spores using time-lapse microscopy. The presented methodology, through the combination of phase-contrast or fluorescence microscopy with image analysis, allows for the generation of large datasets pertinent to different aspects of individual cell microbial behavior and, in tandem with stochastic modeling approaches, constitutes a rather valuable tool in quantitative microbiological risk assessment.

Key words Individual bacterial cells, Individual fungal spores, Time-lapse microscopy, Variability, Growth, Inactivation, Stochastic modeling

1 Introduction

Predictive models have been traditionally developed in the framework of a deterministic approach, describing the behavior of microbial populations in their entirety and providing point estimates of microbial concentrations, without considering individual cell behavior and heterogeneity. The introduction, however, of quantitative microbiological risk assessment (QMRA) in food safety management, a landmark in the history of predictive microbiology, highlighted the importance of developing models capable of predicting the probability distribution of microbial concentrations, prompting the onset of the so-called stochastic predictive microbiology [1–3]. In this framework, the quantification and integration of different sources of variability through stochastic modeling has been tackled in several studies the last 15 years, with individual cell heterogeneity being identified as a rather important source of biological

Verônica Ortiz Alvarenga (ed.), *Basic Protocols in Predictive Food Microbiology*, Methods and Protocols in Food Science, https://doi.org/10.1007/978-1-0716-3413-4_7,

variability in QMRA [1, 3–7]. Beyond quantification of variability, individual cell-based models are anticipated to contribute to the elucidation and better understanding of microbial behavior, with the latter including germination/outgrowth of fungal spores [8–10] and bacterial endospores [11, 12], growth of vegetative cells [7, 13, 14], as well as microbial survival and inactivation [15–17].

Due to the technical difficulties frequently encountered when aspiring to monitor the phenotypic behavior of individual microbial cells, there are not many studies reporting on the development and implementation of relevant methodologies. The most commonly practiced methods for data acquisition in this field have been the ones based on the inoculation of culture broth media in microtiter plates with approximately one cell/spore per well and the conductance of optical density measurements at regular time intervals [18–22]. Nonetheless, this indirect methodological approach presents certain limitations, with the most important ones being (i) the time required for an observation level to be reached (particularly at environmental conditions approaching growth boundaries) and (ii) the method's incompetency to provide information regarding the behavior of immobilized microbial cells within colonies on solid media. Such limitations resulted in the quest for direct and time-efficient methods based, among others, on the use of microfluidic and microscopy-based devices [23–26]. Even in the case of these methods, however, individual cells are monitored solely, with the intercellular interactions potentially occurring in the colonial lifestyle of microorganisms (e.g., on/in solid foods) not being taken into account.

The methodology described herein refers to a quantitative experimental protocol that involves time-lapse microscopy and allows for the monitoring of the behavior (i.e., growth, survival, and inactivation) of individual microbial cells, while fungal spores and bacterial endospores can also be studied with regard to their germination and outgrowth. With particular reference to bacterial cells, this method allows for the assessment and quantitative description of the heterogeneity in the growth [7] and inactivation [17] dynamics of microcolonies originating from single cells. For this purpose, phase-contrast microscopy or fluorescence (confocal laser scanning) microscopy coupled with appropriate staining for cell viability is used. The high-quality images acquired as part of the described protocols allow for the estimation of individual cell division times and growth kinetic parameters as well as inactivation times. The individual cell data (division time, germination time, inactivation time) for the members of the studied microbial populations are collected and, subsequently, fitted to a variety of continuous distributions for defining the distribution that best describes them. The best-fitted probability distribution is further used for describing microbial growth or inactivation using simulation techniques such as Monte Carlo simulation. An overview of the described methodology is illustrated in Fig. 1.

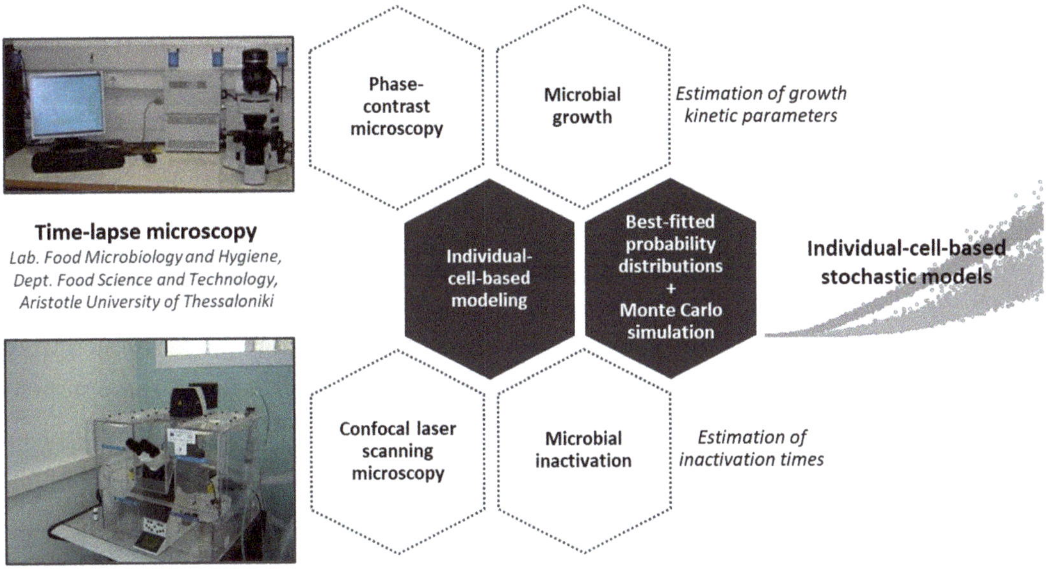

Fig. 1 Schematic of the individual-cell-based modeling approach, using time-lapse microscopy methodologies, described in this chapter

2 Materials

2.1 Laboratory Media

Prepare all culture media according to the manufacturers' instructions and sterilize (autoclaving at 121 °C for 15 min) prior to use.

Standard Growth Media for Bacterial Cultures

1. Tryptone soy broth (TSB; tryptone 17.0 g/L, soy peptone 3.0 g/L, sodium chloride 5.0 g/L, dipotassium phosphate 2.5 g/L, dextrose 2.5 g/L): 30 g of medium's powder is weighed in a Duran bottle and dispersed in 1 L of distilled water. Agitation and warming are applied on a hotplate stirrer to facilitate dissolution.
2. Tryptone soy agar (TSA; tryptone 15.0 g/L, soy peptone 5.0 g/L, sodium chloride 5.0 g/L, agar No. 2 12.0 g/L): 37 g of medium's powder is weighed in a glass conical flask or Duran bottle, dispersed in 1 L of distilled water, and allowed to soak for ca. 10 min. Agitation and warming are applied on a hotplate stirrer to facilitate dissolution. After sterilization, the medium is cooled to 47 °C and thoroughly mixed prior to use.

Standard Culture Medium for Fungal Cultures

1. Malt extract agar (MEA; malt extract 30.0 g/L, mycological peptone 5.0 g/L, agar No. 2 15.0 g/L): 50 g of medium's powder is weighed in a glass conical flask or Duran bottle, dispersed in 1 L of distilled water, and allowed to soak for ca. 10 min. Agitation and warming are applied on a hotplate

stirrer to facilitate dissolution. After sterilization, the medium is cooled to 47 °C, acidified to pH 4.2 with 10% (v/v) lactic acid, and thoroughly mixed prior to use.

Other Materials

1. Ringer's solution (one-quarter strength) as diluent for preparation of microbial suspensions (sodium chloride 2.25 g/L, potassium chloride 0.105 g/L, calcium chloride 0.12 g/L, sodium bicarbonate 0.05 g/L): One tablet (2.5 g/L) is dissolved in 500 mL of deionized water and when completely dissolved is dispensed into containers as required and sterilized.
2. Tween 80 as surfactant for preparation of fungal spore suspensions: 0.1% (v/v) in sterile distilled water.

2.2 Stains

Propidium iodide (1.0 mg/mL in water – filter sterilized) as a cell viability indicator: 120 μL with 2 mL of the inactivation solution (26% NaCl).

2.3 Phase-Contrast Microscope

For the purpose of monitoring bacterial cell and fungal spore kinetic behavior, phase-contrast microscopy can be applied, using a z-motorized optical microscope (Olympus BX61; Olympus, Tokyo, Japan) equipped with a high-resolution device camera (Olympus DP71) [7, 9]. The colonial growth of single cells is monitored using a 100× objective (Olympus) [7], while spore kinetic behavior is monitored at magnification of 40× initially and then at 4× (Olympus) [9]. The samples are maintained at 25 °C using a temperature-controlled stage (Linkam PE60, Linkam Scientific Instruments, Surrey, United Kingdom). The microscope is equipped with a Z stage with a motor controller and an autofocus system, allowing for the movement of the stage to multiple positions. Images of the field of view are captured at 5-min intervals for 6–8 h and 48 h for single cells and spores, respectively, with the lamp of the microscope being automatically turned on and off before and after the capture of an image. High-quality images are provided through the autofocus procedure coupled with an extended depth of focus (EDF) system, allowing for the acquisition of multiple serial images (20–30) in different z-axis planes and then the combination of the best focal areas of the serial images into a single in-focus image (z-stack). Individual images are finally compiled to give a sequence of frames for the field of view, allowing the same cell/spore to be monitored over time throughout the experiment. Based on the compiled image sequence, individual bacterial cells can be monitored during the lag phase and colonial growth (for cells able to initiate growth) [7], while individual fungal spores can be followed during dormancy, germination, germ tube elongation, and finally mycelium formation [9].

2.4 Fluorescence (Confocal Laser Scanning) Microscope

Individual cell growth and, subsequently, inactivation are directly monitored using an inverted confocal laser scanning microscope (TCS – SP5 II; Leica, Heidelberg, Germany) equipped with a SuperZ Galvo stage. The temperature and relative humidity are controlled using Cube 2 and Box for Leica as well as a semi-open stage top incubator for Leica SuperZ Galvo stage rotating insert and condenser.

The formation of microcolonies originating from individual cells is monitored by differential interference contrast (DIC) mode. The sample is maintained under optimum growth conditions (30 °C, 95% RH) for a total duration of 3–4 h till the formation of two-dimensional microcolonies consisting of maximum 100 cells.

Microbial inactivation is monitored using DIC and fluorescence mode simultaneously. Fluorescence of the cells is excited with the 488 nm argon laser line, while samples are observed with a 63× oil immersion objective. Time-lapse observations are performed with Live Data Mode option. Images of the selected field of view are acquired every 5 min using Best Focus tool (contrast based method) with 0.2 μM step size in z axis.

2.5 Image Analysis Software

The growth data generated using the *phase-contrast microscope* are attained using the ImageProPlus image analysis software version 6.3 (Media Cybernetics, Inc., Bethesda, MD, Unites States America). An in-house program developed with the ScopePro module of the software allows for the control of the Z stage as well as for the system's ability to be automatically turned on and off before and after image acquisition. In the case of vegetative cells, data of interest may include cell size, division times, and number of cells in each microcolony (originating from a single cell) with time. Specifically, in high-quality images, cell counting can be performed for up to 100 cells per microcolony, when the growth of the colony is still two-dimensional, using the manual tag of ImageProPlus (*see* **Note 1**). Regarding fungal spores, based on the compiled image sequences and the software's ability to automatically convert pixels into μm (by setting the resolution of the image and the magnification at which it was captured), individual spores can be followed during dormancy, germination, germ tube elongation, and, ultimately, mycelium formation, and corresponding kinetics can be estimated. Since fungal spore germination time can be defined as the time at which the length of the first germ tube is greater than or equal to the greatest dimension of the swollen spore [9, 27], based on the described methodology, an estimate of the distance of the apex hyphae from a swollen spore (until reaching the edge of the optical view) from the point at which the first tube started emerging can be attained.

With reference to the inactivation data collected using the *confocal laser scanning microscope*, a custom MATLAB analysis called Bacterial Single-Cell Analytics (BaSCA) can be used [28]. BaSCA segments and tracks cells in a frame-to-frame manner. This novel methodology yields quantitative expression data of cell lineages, which can illustrate dynamic expression profiles and facilitate mathematical and computational modeling. The obtained sequences are reviewed and analyzed for estimating the PI coverage percentage, i.e., the percentage of the cell surface covered with fluorescent pixels. During exposure, the PI coverage for each cell increases with time in such a way that can be described by a sigmoidal model. After fitting the modified Gompertz equation [29] to PI coverage data for each single cell with time, the lag phase duration is obtained as the individual cell time of inactivation. It has been demonstrated [27] that the 6.6% of the cell surface is covered with PI at the time that corresponds to the geometrical lag time. This low level of coverage corresponding to cell death (in most of the times) coincides with the visual identification of "color" change of the cells.

3 Methods

3.1 Phase-Contrast Microscopy

1. Pre-culture of the microorganism under study:
   - Activate the bacterial strain(s) from working culture(s) on TSA slant(s) into 10 mL of TSB, and incubate at optimum growth temperature for 24 h (*see* **Note 2**).
   - Prepare the spore suspension(s) from heavily sporulating fungal culture(s) in sterile distilled water containing Tween 80 according to the procedure described by Gougouli and Koutsoumanis [9].
2. Perform serial decimal dilutions in Ringer's solution to yield an inoculum level of approximately 10^7 cells or spores per milliliter.
3. Inoculate the agar media with microbial suspension:
   - Add a 20 μL aliquot of the appropriate dilution of bacterial suspension on 500 μL of TSA solidified on a glass slide in a biological safety cabinet, and wait to dry for 5 min (*see* **Notes 1** and **3**); cover the inoculated agar by a coverslip, and seal with silicone to minimize desiccation of the medium (*see* **Note 4**).
   - Add a 100 μL aliquot of the appropriate dilution of fungal spore suspension in a defined area (diameter = 1 mm) on the surface of air-permeable polyethylene plastic film in order to obtain one spore on this area (*see* **Note 1**), and wait to dry for 30 min in a biological safety cabinet; cover

the inoculated area with 1 mL of molten MEA and leave to solidify for 1 min; turn the film upside down and place aseptically stretching it over a lid of a Petri dish (diameter = 35 mm) containing peripherally hydrated cotton to minimize desiccation of the medium (*see* **Note 4**); place the inoculated area in the center of the lid to facilitate microscopic observation.

4. Quickly transfer the prepared sample (slide or Petri dish) from the biological safety cabinet to the microscope stage, and initiate the monitoring as soon as possible (*see* **Note 5**).

3.2 Fluorescence Microscopy

Preparation of the Population to Be Exposed (Single Cells, 2D Microcolonies, 3D Microcolonies)

1. Activate the inoculum by transferring a loopful from TSA slants into 10 mL of TSB without dextrose, and incubate at 37 °C for 16 h.
2. Surface plate 200 μL aliquot of the 16-h culture, after two serial decimal dilutions in Ringer's solution, on 10 mL freshly prepared TSA solidified (approximately 1 h) in a 90 mM Petri dish.
3. Let the inoculated agar to dry in a biological safety cabinet for 10 min, and then remove a disc (of 10 mM diameter) with a scalpel (with the aid of a preformed metallic disc), invert, and place it in a glass bottom dish (0.17 mM thickness; WillCo Wells BV, Amsterdam, The Netherlands) for microscopic observation (*see* **Note 6**).
4. Incubate at 37 °C (RH 95%) for 3–4 h for 2D colonies (less than 100 cells) or approximately 10–12 h for 3D colonies (*see* **Note 2**). The sample is ready for subsequent inactivation steps if single cells are needed.

For survival conditions approaching inactivation boundary, the stress factor can be incorporated in the solid laboratory medium (i.e., pH, salt, etc.) or externally applied during the growth phase (Step 2**). **Step 4** is omitted; proceed to **Step 5**.

Inactivation Step

5. Add 2 mL of the inactivation solution into the glass bottom dish to fully cover the agar disc (*see* **Note** 7).
6. Monitor microbial inactivation using DIC and fluorescence mode simultaneously. Time-lapse observations should be performed with images of the selected field of view being acquired every 5 min (*see* **Note 8**).

3.3 Data Analysis

The individual cell data (division time, germination time, inactivation time) for the members of a microbial population are collected and, subsequently, fitted to a variety of continuous distributions for defining the distribution that best describes them. Depending on the case (growth, inactivation, etc.), different approaches can be employed based on the equation linking the input and the output. After assigning the distribution for the source of variability, techniques such as Monte Carlo simulation can be used to assess the output distribution.

Growth: Kinetic Analysis From growth experimentations and the kinetic analysis, a fitting approach for each microcolony can be adopted. After counting, data are transformed to the respective growth curves showing the exact number of cells in each microcolony originating from a single cell over time. The obtained growth curves are then fitted to an appropriate primary model (e.g., the model of Baranyi and Roberts [30]) for the estimation of lag time (λ) and maximum specific growth rate (μ_{max}). The data of λ and μ_{max} can be fitted to various distributions using a simulation software, e.g., @Risk 6.1 for Excel software (Palisade Corporation, Newfield, United States America). The best-fitted distributions can be further introduced into an exponential model with lag to describe the growth of individual cells using Monte Carlo simulations.

Growth: Division Times From growth experimentations, the division times (first, second, third, etc.) of single cells can be estimated, fitted to various distributions using the simulation software. The best-fitted distributions can be further employed for developing a birth model.

Inactivation: Inactivation Times From inactivation experimentations, the inactivation times of single cells can be fitted to various distributions using the simulation software. The best-fitted distribution can be further used to predict the inactivation of populations using Monte Carlo simulation as described by Aspridou and Koutsoumanis [17].

Simulation analysis can be performed in all the cases described above for various initial population levels. Representative Monte Carlo simulation results for growth and inactivation of *Salmonella enterica* populations are illustrated in Fig. 2.

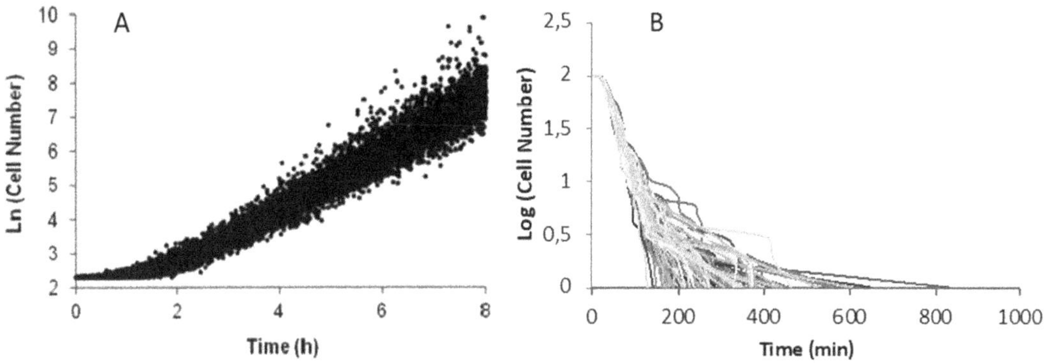

Fig. 2 Monte Carlo simulation results of the stochastic models for the colonial growth (**a**) and inactivation (**b**) of *Salmonella enterica*. (**a**) Model prediction for the growth of a population initially consisting of 10 cells. (Adapted from Ref. [7], Copyright © American Society for Microbiology) and (**b**) model prediction for the inactivation of a population initially consisting of 100 cells. (Adapted from Ref. [17], Copyright 2021, with permission from Elsevier)

4 Notes

1. Be as precise as possible when performing the tasks described in Subheading 3.1; the final cell/spore concentration (after the appropriate dilutions) and the volume and homogeneity of the inoculum on the surface of the solidified agar are of critical importance with regard to the number of individual cells/ spores (and the distance among them) attained in the field of view.
2. The duration of the incubation depends on the incubation conditions and the studied microorganisms. The duration may be different from the presented one.
3. Immediately after inoculation, move in a slight circular mode the slide with the solidified agar to allow for a homogeneous distribution of the added inoculum on its surface; avoid using a bent glass/metal/plastic rod for inoculum spread (potentially affecting the surface integrity of the agar medium and, thus, allowing for cells' internalization inside the medium).
4. Adjust as needed based on the specific experimental protocol that you use, to minimize the desiccation of the medium for the period of the measurements. Perform preliminary experiments involving water activity measurements of the agar medium used to modify (if needed) the described procedure, also keeping in mind that the thickness of the solidified medium cannot exceed a certain level that will hinder microscopic observation.
5. Be aware that the fast initiation of cell/spore microscopic monitoring can be rather challenging; the required time can be variable (up to 5 min for vegetative cells and 10–20 min for

spores) due to the difficulty of manually finding approximately ten well-distanced bacterial cells or one fungal spore in the field of view.

6. Pay attention when transferring the agar disc in order not to damage it. Do not let air bubbles to be trapped between the dish and the agar disc.
7. Add the solution slowly to avoid flushing of the cells, probably by pipetting it. You may prepare two or three samples for being sure of a successful experiment.
8. Pay attention to the focus. Best focus tools can be employed depending on the system.

Beyond the abovementioned notes, a plethora of parameters may determine the fate of a time-lapse microscopy experiment. A non-exhaustive list of common issues and tips that researchers should keep in mind are presented in Table 1.

Table 1
Tips for quantitative time-lapse microscopy

	Issue of concern	Keep in mind	Trouble
Sample	Cell culture	Specific microorganism requirements Density	Abnormal behavior
	Medium	Composition, pH, freshness, fluorescent compounds, antibiotic concentration, solid and/or liquid type, thickness of the substrate if solid	Abnormal behavior, growth defect, increased background
	Chamber	Depending on the medium, microscope stage, multiple imaging, access during the experiment, coatings promoting or preventing cell adherence, working distance of the objective, thickness of the bottom	Leakage of the sample, contamination, dehydration, spherical aberration
	Temperature	Specific needs of the microorganism and the experiment Temperature equilibration	Growth defects, variation of fluorescence level, focus drift
	Atmosphere-sealing	Specific needs of the microorganism Humidification	Abnormal behavior, dehydration of the sample, lateral focus drift
Microscope	Microscope stand	Vibration isolation table Ability to use culture dishes, access to the sample	Vibration from even small activity in the area
	Objective	Numerical aperture	Phototoxicity, photobleaching
	Laser power	Fluctuation	Difficulty in fluorescence data comparison

(continued)

Table 1 (continued)

	Issue of concern	Keep in mind	Trouble
Experiment	Time course Imaging interval	Timescale of the biological phenomenon	Inaccurate description of the phenomenon, photobleaching
	Light exposure	Laser power, scan speed, averaging line or frame, autofocus with transmitted light	Photobleaching
	Pinhole	Capturing efficiency	Loss of in-focus light
	Lateral focus	Size of the solid substrate, sealing	"Moving" object of interest
	Axial focus	Temperature, stable positioning of the culture disc, immobilization of the sample, autofocus range	Blurry – Non-informative images
	Background	Freshness of the solid substrate, room lights, autofluorescent compounds	Difficulty in detection

Adapted from Ref. [6]

References

1. Koutsoumanis KP, Lianou A, Gougouli M (2016) Latest developments in foodborne pathogens modeling. Curr Opin Food Sci 8: 89–98. https://doi.org/10.1016/j.cofs.2016.04.006
2. Nicolaï BM, Van Impe JF (1996) Predictive food microbiology: a probabilistic approach. Math Comput Simul 42:287–292. https://doi.org/10.1016/0378-4754(95)00129-8
3. Voysey PA, Brown M (2000) Microbiological risk assessment: a new approach to food safety control. Int J Food Microbiol 58:173–179. https://doi.org/10.1016/s0168-1605(00)00271-3
4. Aspridou Z, Koutsoumanis KP (2015) Individual cell heterogeneity as variability source in population dynamics of microbial inactivation. Food Microbiol 45:216–221. https://doi.org/10.1016/j.fm.2014.04.008
5. Augustin J-C, Bergis H, Midelet-Bourdin G, Cornu M, Couvert O, Denis C, Huchet V, Lemonnier S, Pinon A, Vialette M, Zuliani V, Stahl V (2011) Design of challenge testing experiments to assess the variability of *Listeria monocytogenes* growth in foods. Food Microbiol 28:746–754. https://doi.org/10.1016/j.fm.2010.05.028
6. Koutsoumanis KP, Aspridou A (2017) Individual cell heterogeneity in predictive food microbiology: challenges in predicting a "noisy" world. Int J Food Microbiol 240:3–10. https://doi.org/10.1016/j.ijfoodmicro.2016.06.021
7. Koutsoumanis KP, Lianou A (2013) Stochasticity in colonial growth dynamics of individual bacterial cells. Appl Environ Microbiol 79: 2294–2301. https://doi.org/10.1128/AEM.03629-12
8. Dagnas S, Gougouli M, Onno B, Koutsoumanis KP, Membré J-M (2015) Modeling red cabbage seed extract effect on *Penicillium corylophilum*: relationship between germination time, individual and population lag time. Int J Food Microbiol 211:86–94. https://doi.org/10.1016/j.ijfoodmicro.2015.07.008
9. Gougouli M, Koutsoumanis KP (2013) Relation between germination and mycelium growth of individual fungal spores. Int J Food Microbiol 161:231–239. https://doi.org/10.1016/j.ijfoodmicro.2012.12.006
10. Peleg M, Normand MD (2013) Modeling of fungal and bacterial spore germination under static and dynamic conditions. Appl Environ Microbiol 79:6765–6775. https://doi.org/10.1128/AEM.02521-13
11. Kakagianni M, Aguirre JS, Lianou A, Koutsoumanis KP (2017) Effect of storage temperature on the lag time of *Geobacillus stearothermophilus* individual spores. Food Microbiol 67:76–84. https://doi.org/10.1016/j.fm.2017.04.009
12. Williams B, López-García M, Gillard JJ, Laws TR, Lythe G, Carruthers J, Finnie T, Molina-

París C (2021) A stochastic intracellular model of anthrax infection with spore germination heterogeneity. Frontiers Immunol 12: 688257. https://doi.org/10.3389/fimmu.2021.688257

13. D'Souza GG, Povolo VR, Keegstra JM, Stocker R, Ackermann M (2021) Nutrient complexity triggers transitions between solitary and colonial growth in bacterial populations. ISME J 15:2614–2626. https://doi.org/10.1038/s41396-021-00953-7
14. Fritsch L, Baleswaran A, Bergis H, Lintz A, Hamon E, Stahl V, Augustin JC, Guillier L (2021) A microscopy-based approach for determining growth probability and lag time of individual bacterial cells. Food Res Int 140: 110052. https://doi.org/10.1016/j.foodres.2020.110052
15. Abe H, Koyama K, Kawamura S, Koseki S (2019) Stochastic modeling of variability in survival behavior of *Bacillus simplex* spore population during isothermal inactivation at the single cell level using a Monte Carlo simulation. Food Microbiol 82:436–444. https://doi.org/10.1016/j.fm.2019.03.005
16. Aspridou Z, Akritidou T, Koutsoumanis KP (2018) Simultaneous growth, survival and death: the trimodal behavior of *Salmonella* cells under osmotic stress giving rise to "Phoenix phenomenon". Int J Food Microbiol 285: 103–109. https://doi.org/10.1016/j.ijfoodmicro.2018.07.012
17. Aspridou Z, Balomenos A, Tsakanikas P, Manolakos E, Koutsoumanis K (2019) Heterogeneity of single cell inactivation: assessment of the individual cell time to death and implications in population behavior. Food Microbiol 80:85–92. https://doi.org/10.1016/j.fm.2018.12.011
18. Pin C, Baranyi J (2006) Kinetics of single cells: observation and modelling of a stochastic process. Appl Environ Microbiol 72:2163–2169. https://doi.org/10.1128/AEM.72.3.2163-2169.2006
19. Baranyi J (2002) Stochastic modelling of bacterial lag phase. Int J Food Microbiol 73:203–206. https://doi.org/10.1016/s0168-1605(01)00650-x
20. Blana VA, Lianou A, Nychas GJ (2015) Assessment of the effect of a *Salmonella enterica* ser. Typhimurium culture supernatant on the single-cell lag time of foodborne pathogens. Int J Food Microbiol 215:143–148. https://doi.org/10.1016/j.ijfoodmicro.2015.09.009
21. Métris A, Le Marc Y, Elfwing A, Ballagi A, Baranyi J (2005) Modelling the variability of lag times and the first generation times of single cells of *E. coli*. Int J Food Microbiol 100:13–19. https://doi.org/10.1016/j.ijfoodmicro.2004.10.004
22. Smelt JP, Otten GD, Bos AP (2002) Modelling the effect of sublethal injury on the distribution of the lag times of individual cells of *Lactobacillus plantarum*. Int J Food Microbiol 73(2–3):207–212. https://doi.org/10.1016/s0168-1605(01)00651-1
23. Elfwing A, LeMarc Y, Baranyi J, Ballagi A (2004) Observing growth and division of large numbers of individual bacteria by image analysis. Appl Environ Microbiol 70:675–678. https://doi.org/10.1128/AEM.70.2.675-678.2004
24. Niven GW, Fuks T, Morton JS, Rua SA, Mackey BM (2006) A novel method for measuring lag times in division of individual bacterial cells using image analysis. J Microbiol Methods 65:311–317. https://doi.org/10.1016/j.mimet.2005.08.006
25. Siegal-Gaskins D, Crosson S (2008) Tightly-regulated and heritable division control in single bacterial cells. Biophys J 95:2063–2072. https://doi.org/10.1529/biophysj.108.128785
26. Wakamoto Y, Ramsden J, Yasuda K (2005) Single-cell growth and division dynamics showing epigenetic correlations. Analyst 130: 311–317. https://doi.org/10.1039/b409860a
27. Gougouli M, Koutsoumanis KP (2012) Modeling germination of fungal spores at constant and fluctuating temperature conditions. Int J Food Microbiol 152:153–161. https://doi.org/10.1016/j.ijfoodmicro.2011.07.030
28. Balomenos AD, Tsakanikas P, Aspridou Z, Tampakaki AP, Koutsoumanis KP, Manolakos ES (2017) Image analysis driven single-cell analytics for systems microbiology. BMC Syst Biol 11:43. https://doi.org/10.1186/s12918-017-0399-z
29. Zwietering MH, Jongenburger I, Rombouts FM, Van't Riet K (1990) Modeling of the bacterial growth curve. Appl Environ Microbiol 56:1875–1881. https://doi.org/10.1128/aem.56.6.1875-1881.1990
30. Baranyi J, Roberts TA (1994) A dynamic approach to predicting bacterial growth in food. Int J Food Microbiol 23:277–294. https://doi.org/10.1016/0168-1605(94)90157-0

Chapter 8

Dynamic Models for Predictive Microbiology

Daniel Angelo Longhi, Bruno Augusto Mattar Carciofi, Gláucia Maria Falcão de Aragão, and João Borges Laurindo

Abstract

First-order ordinary differential equations have been used as mathematical models to describe and predict the dynamics of foodborne microorganisms. The most suitable models are based on an elementary block that quantifies the instantaneous rate of microbial growth (positive) or inactivation (negative) as directly proportional to the instantaneous microbial concentration. Adjustment functions multiply this elementary block to describe deviations from the log-linear behavior, leading to the adaptation (lag) and stationary phases for growth or the shoulder and tail phases for inactivation. Parameter estimation is performed from experimental data obtained under constant or variable extrinsic and intrinsic conditions. The latter can be performed by the traditional two-step (adjusting primary and then secondary equations) or one-step (fitting coupled primary-secondary equations) approaches. This chapter details useful procedures applied to quantitatively model the growth and inactivation dynamics of microorganisms in food, including some case studies of the microbial dynamics under constant and variable conditions.

Key words Microbial growth, Microbial inactivation, Food safety, Food quality, Primary model, Secondary model, Parameter estimation

1 Introduction

The study of the microbial population dynamics has long been a topic explored by researchers. The first modes of the dynamics of microbial inactivation in foods were conducted in the early twentieth century [1]. Microbial dynamics were well understood for intrinsic (e.g., pH, water activity, organic acid concentration) and extrinsic (e.g., temperature, gas concentration) constant conditions. In the last decades, first-order ordinary differential equations have been used to describe the microbial dynamics for factors that vary over time [2, 3]. Since then, the mathematical models began to describe the microbial dynamics applied to food processing and storage under constant and variable conditions.

The main mathematical models used to describe the microbial dynamics are based on an elementary block that quantifies the

Verônica Ortiz Alvarenga (ed.), *Basic Protocols in Predictive Food Microbiology*, Methods and Protocols in Food Science, https://doi.org/10.1007/978-1-0716-3413-4_8,

change in microbial concentration N (CFU/(g or mL)) over time t (min or h) through a first-order differential equation [4], as shown in Eq. 1. These models that describe microbial kinetics are known as primary models [5]. In Eq. 1, r (1/min or 1/h) is the specific rate of the microbial species, which depends on the environmental conditions (*cond*) to which the species is subjected in addition to the species' phenotypic factors.

$$\frac{dN(t)}{dt} = r(t, N, \text{cond})N(t) \tag{1}$$

Equation 1 can be modified algebraically, resulting in Eq. 2, in which the dependent variable becomes the logarithm of the microbial concentration $y(t) = \ln N(t)$.

$$\frac{1}{N(t)}\frac{dN(t)}{dt} = \frac{d\ln N(t)}{dt} = \frac{dy(t)}{dt} = r(t, N, \text{cond}) \tag{2}$$

The elementary block presented in Eq. 2 is a basis for many mathematical models describing microbial growth and inactivation dynamics. The mathematical difference between growth and inactivation is the sign of the specific rate (r) since its positive value results in the representation of microbial growth (generally r is represented by μ), and its negative value results in inactivation (generally r is represented by k). Eq. 2 is a linear relationship between y and t. However, very often, experimental data on microbial dynamics shows deviations of this behavior, mostly at the beginning and end of the observed kinetics, even under constant conditions.

Thus, shape functions specific for growth or inactivation multiply the elementary block to represent changes in the dynamics. These functions can depend on time, concentration, and environmental parameters, assuming values from zero to one. When a function value reaches the value one, that function does not interfere with the microbial dynamics. On the other hand, when a function value is less than one, that function influences the microbial dynamics, indicating that the microorganism is not growing or inactivating at the maximum rate.

For microbial growth, functions multiplying the elementary block describe the adaptation (f_{lag}) and stationary (f_{sta}) phases, according to Eq. 3. Therefore, Eq. 3 can result in four main formats – (i) pure exponential growth, without lag and stationary phases ($f_{\text{lag}} = f_{\text{sta}} = 1$); (ii) exponential growth with lag phase ($0 < f_{\text{lag}} \leq 1$) and without stationary phase ($f_{\text{sta}} = 1$); (iii) exponential growth without lag phase ($f_{\text{lag}} = 1$) and with the stationary phase ($0 < f_{\text{sta}} \leq 1$); and (iv) exponential growth with lag phase ($0 < f_{\text{lag}} \leq 1$) and stationary phase ($0 < f_{\text{sta}} \leq 1$) – which results in the well-known sigmoidal growth curve.

$$\frac{dy(t)}{dt} = \mu_{\max} f_{\text{lag}} f_{\text{sta}} \tag{3}$$

For microbial inactivation, functions multiply the elementary block to describe a shoulder (f_{sho}) and tail (f_{tai}) phases, according to Eq. 4. The parameter n is also added as an exponent of $k_{\max}$ to describe the cases of nonlinear inactivation, and a subscript i may be used to specify subpopulations (or distinct species). Therefore, Eq. 4 can describe, at least, eight main formats: (i) linear inactivation ($n = 1$) of one population ($i = 1$), without shoulder and tail ($f_{\text{sho}} = f_{\text{tai}} = 1$); (ii) linear inactivation ($n = 1$) of one population ($i = 1$), with shoulder ($0 < f_{\text{sho}} \leq 1$) and without tail ($f_{\text{tai}} = 1$); (iii) linear inactivation ($n = 1$) of one population ($i = 1$), without shoulder ($f_{sho} = 1$) and with tail ($0 < f_{\text{tai}} \leq 1$); (iv) linear inactivation ($n = 1$) of one population ($i = 1$), with shoulder ($0 < f_{\text{sho}} \leq 1$) and tail ($0 < f_{\text{tai}} \leq 1$); (v) nonlinear inactivation with convex curvature ($n > 1$) of one population ($i = 1$), without shoulder and tail ($f_{\text{sho}} = f_{\text{tai}} = 1$); (vi) nonlinear inactivation with concave curvature ($n < 1$) of one population ($i = 1$), without shoulder and tail ($f_{\text{sho}} = f_{\text{tai}} = 1$); (vii) linear inactivation ($n = 1$) of two subpopulations ($i = 1$ and 2), without shoulder and tail ($f_{\text{sho}} = f_{\text{tai}} = 1$); and (viii) linear inactivation ($n = 1$) of two subpopulations ($i = 1$ and 2), with shoulder ($0 < f_{\text{sho}} \leq 1$) and without tail ($f_{\text{tai}} = 1$). For formats (vii) and (viii), the total population at a certain time ($N_t(t)$) is calculated by summing the two subpopulations at that time ($N_1(t)$ and $N_2(t)$, i.e., $N_t(t) = N_1(t) + N_2(t)$) or their logarithmic transformations ($y_i(t) = \ln(N_i(t))$), according to Eq. 5.

$$\frac{dy_i(t)}{dt} = -nk^n_{\max,i} t^{n-1} f_{\text{sho}} f_{\text{tai}} \tag{4}$$

$$y_t(t) = \ln\left(e^{y_1(t)} + e^{y_2(t)}\right) \tag{5}$$

Many mathematical models can be found in the literature to describe the dynamics of microbial growth [6–8] and inactivation [8, 9]. One of the mathematical models most used to describe microbial growth is the Baranyi and Roberts model [3], presented in Eqs. 6 and 7, in which $Q(t)$ is a function that is related to the physiological state of the cells and $y_{\max}$ is the logarithm of the maximum microbial concentration ($y_{\max} = \ln N_{\max}$).

$$\frac{dy(t)}{dt} = \mu_{\max}\left(\frac{1}{1 + e^{-Q(t)}}\right)\left(1 - e^{y(t) - y_{\max}}\right) \tag{6}$$

$$\frac{dQ(t)}{dt} = \mu_{\max} \tag{7}$$

The comparison of Eq. 6 with Eq. 3 allows one to conclude that, for the Baranyi and Roberts model, the functions multiplying μ_{max} describe the adaptation phase ($f_{\text{lag}} = \left(\frac{1}{1+e^{-Q(t)}}\right)$) and the

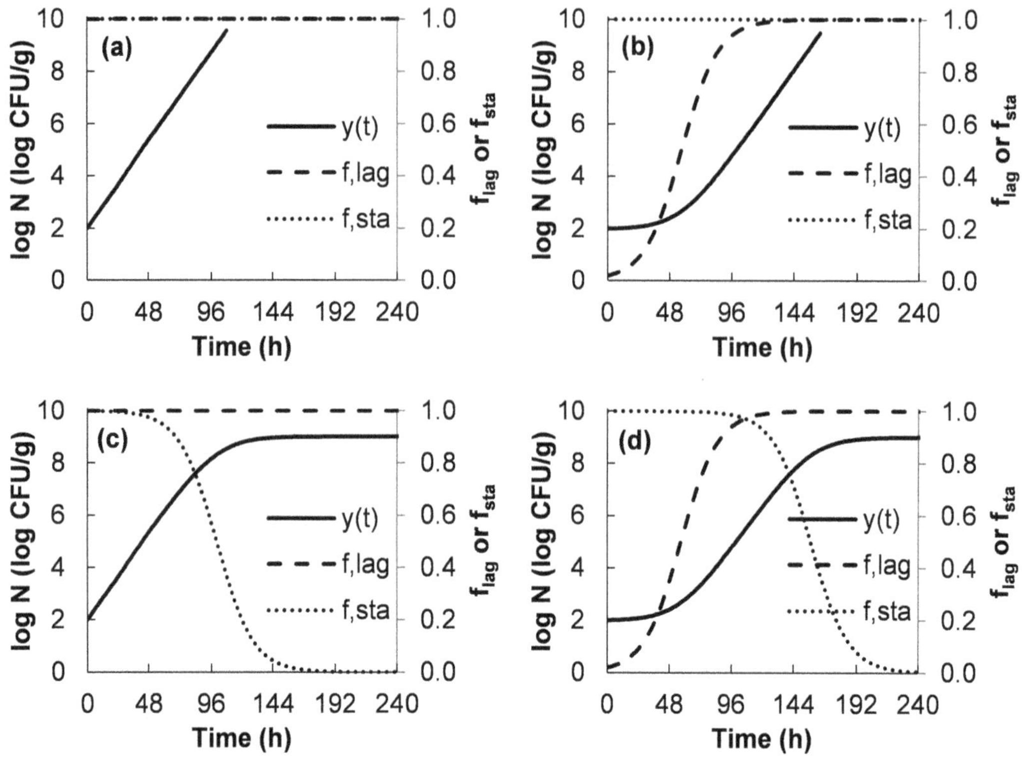

Fig. 1 Simulations of microbial growth dynamics with constant factors in its four main formats: (**a**) pure exponential growth, without lag and stationary phases; (**b**) exponential growth with lag phase and without stationary phase; (**c**) exponential growth without lag phase and with stationary phase; and (**d**) exponential growth with lag and stationary phases. The curves were generated with the Baranyi and Roberts model, Eqs. 6 and 7, with $y_0 = 2$ log (CFU/g); $\mu_{max} = 0.07$ h^{-1}; $Q_0 = -4$ (**b** and **d**); and $y_{max} = 9$ log (CFU/g) (**c** and **d**). f_{lag} is the function of the lag phase and f_{sta} is the function of the stationary phase

stationary phase ($f_{sta} = \left(1 - e^{y(t) - y_{max}}\right)$) of microbial growth. Other mathematical models differ from this one in proposing alternatives for both functions.

Equations 6 and 7 solution results in a sigmoid curve with adaptation, exponential, and stationary phases. The exclusion of functions describing the adaptation phase, stationary phase, or both will result in a simplified curve. Simulations of the resulting curves with different considerations in the model are shown in Fig. 1.

The Baranyi and Roberts model with modifications, as shown in Eqs. 8 and 9, can also be used for describing the microbial inactivation [10], in which $R(t)$ is a variable related to the physiological state of the cells in the inactivation process ($R(t) = \log C_c(t)$) and y_{res} is the logarithm of the residual microbial concentration ($y_{res} = \ln N_{res}$). Analogously to Eq. 6, the functions that multiply k_{max} in Eq. 8 describe the shoulder and tail of the microbial inactivation curve. Therefore, Eqs. 8 and 9 solution results in a sigmoid curve with shoulder, log-linear inactivation, and tail. As described for Eqs. 6 and 7, Eqs. 8 and 9 solution without the function for shoulder, tail, or both will result in a simplified curve with the correspondent exclusion performed.

$$\frac{dy(t)}{dt} = -k_{\max}\left(\frac{1}{1+e^{-R(t)}}\right)\left(1-e^{y_{\text{res}}-y(t)}\right) \tag{8}$$

$$\frac{dR(t)}{dt} = -k_{\max} \tag{9}$$

When the intrinsic and extrinsic factors of the food are considered constant, then the parameters of the previous equations have constant values. In this case, the ordinary differential equations can be solved algebraically; thus, the model can be expressed by an explicit function. On the other hand, when the intrinsic and extrinsic factors are variable, then the parameters of the equations are not constant. In this case, the ordinary differential equations can be solved numerically because the dynamic model depends on other models, known as secondary models [5], to describe the dependence of primary model parameters with the variable intrinsic and extrinsic factors.

Secondary models do not have an elementary block as primary models do. Instead, the secondary models can be any equations that represent the dispersion of parameters as a function of variable factors. Some secondary models have been proposed and widely used in specific situations; some examples are presented in [8].

In the context of the briefly given introduction, this chapter aims to show a frequent sequence of procedures applied to model the growth and inactivation dynamics of microorganisms in food, including some demonstrative case studies of the microbial dynamics under constant and variable conditions.

2 Materials

Experimental kinetic data of the behavior of the target microorganism in the food of interest under certain conditions are the minimum material for the mathematical modeling of the microbial dynamics. In this context, the modeler who does not have any experimental data can resort to experimental databases (e.g., the Combined Database for Predictive Microbiology – ComBase [11]), request data presented in scientific articles from their authors, or extract data from reliable and available sources with the help of software (e.g., WebPlotDigitizer [12]). If the experimental data are insufficient or unavailable, the modeler can still simulate microbial behavior with mathematical models and hypothetical parameters, but the results will be theoretical.

If the researcher can collect original experimental data in the laboratory, the gold standard for microbial quantification in food is the plate counting method. The basic materials needed for microbial quantification in food are established in federal norms, or documents from the International Standardization Organization (ISO) can be consulted and followed. Among the master documents, one can see ISO 7218 [13], which shows general requirements and guidance for microbiological examinations, and ISO 11133 [14], which shows standards for preparation, production, storage, and performance testing of culture media.

The quantification and recording of the values of the intrinsic and extrinsic conditions of the culture medium or food under analysis are fundamental, as they govern the microbial dynamics. Systems for recording the temperature evolution of the food and ambient are essential, as the temperature is usually the factor with high variation during food processing, distribution, and storage. Also, data on the environment's pH, water activity, and air composition is required for more exact parameter estimation and model prediction. ISO 20976-1 [15] suggests that the microbiology laboratory should have the equipment to prepare the test portions, store them under suitable conditions, or monitor how their characteristics change during the test study. These include apparatus for packaging, chilled incubator, climate-control chamber, pH meter, headspace gas analyzer, and loggers for measuring temperature and relative humidity storage conditions of the test unit.

To estimate the parameters of a primary model, the theoretical minimum number of experimental data needed in each kinetics is equal to the number of model parameters plus one. However, the uncertainty of the microbiological data can lead to estimated parameters with wide confidence intervals. Therefore, even if it is possible to estimate the model parameters with a minimum number of data, it is recommended that more data than the minimum should be collected, and the data must be sufficient to represent all the kinetic behavior (growth and inactivation phases).

The analogy can be applied to estimate the parameters of the secondary models, in which the theoretical minimum number of experimental kinetics needed is equal to the number of model parameters plus one. Likewise, even if it is possible to estimate the model parameters with a minimum number of experimental kinetics, it is recommended to perform more trials than the minimum, regarding the full dependence of the parameters with the intrinsic and extrinsic factors analyzed.

With the experimental data available, another fundamental material is a computer with good processing capacity and numerical calculation software. Several software options are available, free or proprietary (e.g., Matlab, R, Microsoft Office, Libre Office). The choice of the software depends on the features it provides in modeling tasks, and there is software that can stand out for its ease of use, available toolboxes, or user-friendly interface. This chapter uses Matlab R2020b software (Mathworks, Natick, USA).

3 Methods

3.1 Parameter Estimation

The traditional approach for parameter estimation is performed in two steps, which are applied to experimental data obtained under constant conditions. In the first step, the primary model is fitted to each of the experimental kinetic datasets, and then the parameters of the primary model under the measured conditions are estimated. The values of the different parameters of the primary models (e.g., $\mu_{\max}$, λ, y_0, and $y_{\max}$ for growth) in each constant conditions are dispersed as a function of each measured intrinsic or extrinsic factor to observe trends. Then, appropriate secondary models are fitted to the data to describe the dependence of each parameter with each condition, and their parameters are estimated.

Another approach for parameter estimation is performed in one step, which can be applied to data obtained under constant or variable conditions. In this single step, the primary and secondary models are fitted simultaneously to all the available experimental data, resulting in the estimation of all the parameters of the primary and secondary models.

Good alternatives for obtaining the microbial growth and inactivation model parameters are to consult the literature and the microbiological databases (e.g., ComBase). These options require much caution, in which all information must be verified. An essential piece of information, in addition to the microbial species of interest, is the characterization of the strain(s) (origin, the form of isolation, identification, etc.), as different strains may have different behavior, which is normally attributed to the variability of the species. Other essential information about the type of food in which the parameters were estimated is the characterization of the intrinsic and extrinsic factors of the experiment. When the information obtained is similar to the characteristics of the study in progress, it is possible to design a reduced number of experiments to validate these parameters.

Regarding the parameter estimation of the secondary models, when the microbial growth is analyzed at temperatures below the optimum growth temperature of the microorganism (T_{opt}), the dependence of $\mu_{\max}$ with the temperature (T) is frequently well described with the square-root model [16], Eq. 10, in which b ($h^{-0.5}.°C^{-1}$) is an empirical parameter associated with the slope of the square root of μ_{max} as a function of T. When the microbial growth is analyzed over the entire temperature range, from the minimum temperature ($T_{\min}$) to the maximum temperature ($T_{\max}$), the dependence of $\mu_{\max}$ on T is frequently well described with the cardinal parameter model [17], Eq. 11, in which μ_{opt} is the maximum specific growth rate at the optimal temperature (T_{opt}).

$$\sqrt{\mu_{max}} = b(T - T_{min}) \tag{10}$$

$$\mu_{max} = \mu_{opt} \frac{(T - T_{max})(T - T_{min})^2}{(T_{opt} - T_{min})\left[(T_{opt} - T_{min})(T - T_{opt}) - (T_{opt} - T_{max})(T_{opt} + T_{min} - 2T)\right]} \tag{11}$$

3.1.1 Case Study Under Constant Conditions

The case study under constant conditions is based on the two-step modeling approach. For microbial growth under constant conditions, Eqs. 6 and 7 of the Baranyi and Roberts model can be solved analytically [18], and the solution can be presented explicitly by Eqs. 12 and 13, in which h_0 is a transformation of the parameter Q_0, as shown in Eq. 14.

$$y(t) = y_0 + \mu_{max} A(t) - \ln\left(1 + \frac{e^{\mu_{max} A(t)} - 1}{e^{y_{max} - y_0}}\right) \tag{12}$$

$$A(t) = t + \frac{1}{\mu_{max}} \ln\left(e^{-\mu_{max} t} + e^{-h_0} - e^{-\mu_{max} t - h_0}\right) \tag{13}$$

$$h_0 = \ln\left(1 + \frac{1}{e^{Q_0}}\right) \tag{14}$$

A dataset of the *Lactobacillus viridescens* growth in vacuum-packed sliced ham stored at 8 °C [19] was used to illustrate the fitting of the primary model (Eqs. 12 and 13) to the experimental data. At times 0, 6, 24, 30, 48, 54, 72, 78, 96, 102, 120.5, 144.75, 168, 192, 216, 240, 292, 336, and 384 h of incubation, the following average concentration values (duplicate) were measured: 5.08, 4.99, 6.03, 7.65, 8.86, 8.98, 10.52, 11.29, 12.37, 13.47, 15.03, 17.68, 18.61, 19.07, 18.83, 20.14, 21.45, 19.95, and 20.32 ln CFU/g. In the command window of Matlab, the basic inputs of the times (t), the microbial concentrations (C), and the model equation (*model*) should be provided with the following syntax. Note that the text written after the "%" symbol refers to comments added to the computational routine; Matlab does not process it in the code.

```
t = [0; 6; 24; 30; 48; 54; 72; 78; 96; 102; 120.5; 144.75; 168;
192; 216; 240; 292; 336; 384]; % time data
C = [5.08; 4.99; 6.03; 7.65; 8.86; 8.98; 10.52; 11.29; 12.37;
13.47; 15.03; 17.68; 18.61; 19.07; 18.83; 20.14; 21.45; 19.95;
20.32]; % concentration data
model=fittype('y0+mu*(x+(1/mu)*(log(exp(-mu*x)+exp(-h0)-exp
((-mu*x)-h0))))-log((((exp(mu*(x+(1/mu)*(log(exp(-mu*x)+exp
(-h0)-exp((-mu*x)-h0))))))-1)/(exp(ymax-y0)))+1)'); % defini-
tion of the primary model function
```

The fitting of the model to the experimental data can be performed in the command window with the following syntax.

```
[fitting,gof]=fit(t,C,model,'StartPoint',[1 0.1 5 20]) % fit-
ting the model to the data
```

The estimated parameters (coefficients) in this case study were $h_0 = 0.515$, $\mu_{max} = 0.08802$ 1/h, $y_0 = 4.883$ ln CFU/mL, and $y_{max} = 20.04$ ln CFU/mL, and the 95% confidence bounds of the model parameters were (−1.087, 2.117), (0.07756, 0.09847), (3.819, 5.946), and (19.49, 20.59), respectively. The *gof* calculates the goodness-of-fit. In this case study, some of them were the coefficient of determination (R^2) equal to 0.9907, and the root-mean-square errors (RMSE) equal to 0.6043. The following syntax in the Matlab command window shows the experimental data (blue circles) and the primary model fitted to the data (red line) in a figure. This figure is shown in Fig. 2.

```
plot(fitting, 'r', t, C, 'bo', 'fit') % plot the data (blue
circles) and the primary model fit (red curve) in a figure
xlabel('t (h)'); ylabel('y (ln CFU/g)'); % labeling the x and y
axes, respectively
```

A dataset of the maximum specific growth rates of *Lactobacillus viridescens* from 4 to 30 °C [20] was used to illustrate the fitting of the secondary model (Eq. 10) to the data. At temperatures 4, 8, 12, 16, 20, and 30 °C (duplicate), the following values of the maximum specific growth rate were estimated: 0.0178, 0.0258,

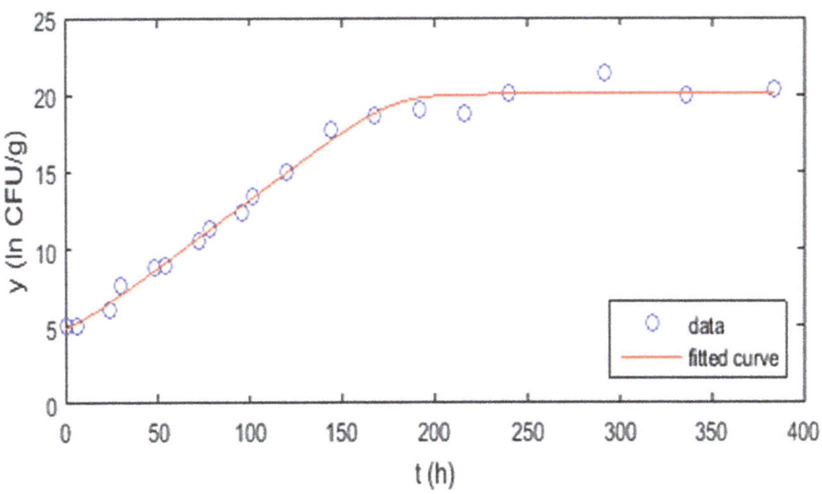

Fig. 2 Dataset of the growth of *Lactobacillus viridescens* in vacuum-packed sliced ham (circles) stored at 8 °C and the fitting of the Baranyi and Roberts primary model (continuous line) to the experimental data generated in Matlab

0.0575, 0.0689, 0.1513, 0.1602, 0.2546, 0.2724, 0.4132, 0.4363, 0.7386, and 0.8254 1/h. In the command window of the Matlab, the basic inputs of the temperatures (*Temp*), the square root of the values of the maximum specific growth rates (*sqr_mu*), and the model equation (*sec_model*) should be provided with the following syntax:

```
Temp = [4; 4; 8; 8; 12; 12; 16; 16; 20; 20; 30; 30]; % Temperature
sqr_mu = sqrt([0.0178; 0.0258; 0.0575; 0.0689; 0.1513; 0.1602; 0.2546; 0.2724; 0.4132; 0.4363; 0.7386; 0.8254]); % square-root of the maximum specific growth rates
sec_model=fittype('b*(x-Tmin)'); % definition of the secondary model function
```

The fitting of the model to the data can be performed in the command window with the following syntax:

```
[sec_fitting,gof]=fit(Temp,sqr_mu,sec_model,'StartPoint',[0.1 0]) % fitting the model to the data
```

The estimated parameters (coefficients) in this case study were $b = 0.029$ 1/($h^{0.5}$.°C) and $T_{min} = -1.329$ °C, and the 95% confidence bounds of the model parameters were (0.02702, 0.03099) and (−2.588, −0.06932), respectively. The coefficient of determination (R^2) was equal to 0.9906, and the root-mean-square errors (RMSE) were equal to 0.0261. The following syntax in the Matlab command window shows the square root of the maximum specific growth rate (blue circles) and the fitted secondary model (red line) in a figure. The figure generated with this procedure is shown in Fig. 3.

```
plot(sec_fitting, 'r', Temp, sqr_mu, 'bo', 'fit') % plot the data (blue circles) and the secondary model fit (red curve) in a figure
xlabel('T (°C)'); ylabel('\surd mu (\surd 1/h)'); % labeling the x and y axes, respectively
```

3.2 Model Simulations to Predict the Microbial Dynamic

When the parameters of the primary and secondary models are already estimated, as presented in Subheading 3.1, it is possible to perform simulations with the mathematical models to predict the dynamics of the microbial behavior under different conditions. It is important to emphasize that the predictions must be restricted to the range of values previously quantified for the factors under analysis. For example, if the experimental kinetic data were obtained for microbial growth at five constant temperatures (4 °C, 8 °C, 12 °C, 16 °C, and 20 °C) and the other factors

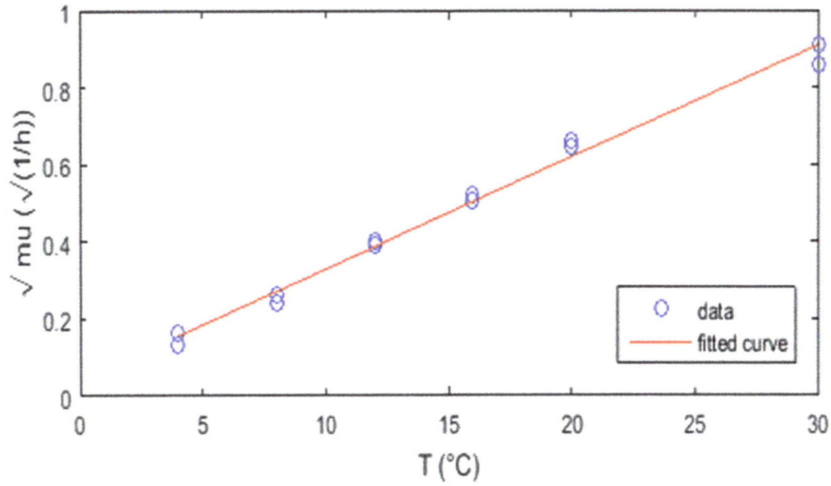

Fig. 3 Dataset of the square root of the maximum specific growth rates of *Lactobacillus viridescens* (circles) at 4, 8, 12, 16, 20, and 30 °C and the square-root secondary model (continuous line) fitted to the experimental data generated in Matlab

(e.g., pH, water activity) were constant at the same value, then the simulations can be performed to predict the dynamics of the microbial growth at any temperature in the range from 4 °C to 20 °C, as well as different temperature oscillations. Simulations of the microbial dynamics should be avoided for cases where the temperature exceeds this range and for different values of other factors (pH and water activity). Extrapolations can result in deviations from the actual behavior of the microorganism.

The simulation of the dynamics of the microbial behavior can be performed using different strategies. As the elementary block that describes the dynamics of the microbial behavior is structured through ordinary differential equations (ODE, e.g., Eqs. 6, 7, 8, and 9), their solution depends on the characteristic of the factors (intrinsic and extrinsic) along the time. When the factors are constant along the time, ODE solutions can be obtained by the explicit functions (e.g., Eqs. 12 and 13). When factors vary over time, ODE solutions are obtained by numerical methods, such as the fourth order Runge-Kutta.

3.2.1 Case Study of Variable Conditions

Two datasets were used to illustrate the predictions of the growth of *L. viridescens* in vacuum-packed sliced ham stored under non-isothermal conditions [19] obtained with the primary model of Baranyi and Roberts, Eqs. 6 and 7, and the secondary square-root model of Ratkowsky and coworkers, Eq. 10. The two datasets show temperature variations from 4 °C to 16 °C, in which the first and second ones have temperature increases and decreases in plateaus, respectively.

In the first dataset, with the ham temperature increasing in plateaus from 4 °C to 16 °C, growth was quantified at times 0, 19, 43, 63, 72, 87, 91.6, 96, 105, 120, 135, 144, 159, and 168 h of incubation with the following mean concentration values (duplicate): 9.50, 10.01, 10.80, 11.62, 12.01, 13.30, 13.22, 13.49, 15.75, 18.40, 18.84, 19.38, 20.00, and 19.09 ln CFU/g. In Matlab, for the first dataset, the basic entries of the time (*tdata*) and the microbial concentration (*Cdata*) should be typed in the command window with the following syntax:

```
tdata = [0; 19; 43; 63; 72; 87; 91.6; 96; 105; 120; 135; 144;
159; 168]; % times of dataset one
Cdata = [9.50; 10.01; 10.80; 11.62; 12.01; 13.30; 13.22;
13.49; 15.75; 18.40; 18.84; 19.38; 20.00; 19.09]; % concentra-
tions of dataset one
```

A new script, an option that appears in the upper left corner of the Matlab, should be opened to define the model with the following syntax. This script should be saved as a ".m" file format with the name "odemodel.m."

```
function dy = odemodel(t,y)
global tp Tp tsize % global temperature and time data
k=1;
for k = 1:k:(tsize(1,2)-1) % attribution of the temperature to
each time - next five lines
if (t >= tp(k) & t <= tp(k+1))
T = Tp(k);
end;
end;
ymax = 21.0; % maximum concentration value of L. viridescens
b = 0.029; Tmin = -1.3; % secondary model parameters of
L. viridescens
mu = (b*(T-Tmin))^2; % square-root secondary model
dy = zeros(2,1);
dy(1) = mu; % ODE Baranyi and Roberts model for dQ/dt
dy(2) = mu*(1/(1+exp(-y(1))))*(1-exp(y(2)-ymax)); % ODE Bar-
anyi and Roberts model for dy/dt
```

After the model file has been saved, in the command window of Matlab, the input information should be typed to specify the temperature profile, the time integration range, and the initial conditions of the differential equations. The definition of the temperature profile is a relevant task that requires attention in programming, as the availability of experimental temperature records is limited to predefined times, according to the characteristic of the datalogger. In turn, the computational routine will need temperature data at unavailable time intervals; therefore, a programming strategy must be employed. For simplicity, in the temperature profile of the dataset, the past recorded temperature value at each absent measure was considered.

```
global tp Tp tsize % global temperature and time data
tp = [0 63.0 63.0 91.6 91.6 105.0 105.0 168.0]; % time for
time-temperature profile one
Tp = [4.8 4.8 8.4 8.4 12.7 12.7 15.8 15.8]; % temperature for
time-temperature profile one
tsim = [0:0.01:168];
tsize = size(tp);
C0 = Cdata(1);
h0 = 0;
Q0 = log(1+(1/exp(h0)));
ysim = [h0 C0];
[tpred,Cpred] = ode23(@odemodel,tsim,ysim);
```

The results of the growth dynamics simulated for the first dataset can be observed through a figure and qualitatively compared with the distribution of experimental values under the same simulation conditions. The following syntax in the Matlab command window shows the experimental data of the microbial concentration (blue circles), the temperature (dashed red line), and the simulated curve resulting from the growth model (continuous black line).

```
figure(1) % open a new figure 1 window
plot(tdata, Cdata, 'bo') % plot the symbols of the experimental
dataset, in blue color
hold all % hold all Information in the figure
plot(tpred, Cpred(:,2), 'k-') % plot the continuous line of the
growth predicted by the model, in black color
plot(tp, Tp, 'r--') % plot the dashed line of the time-
temperature profile, in red color
xlabel('t (h)'); ylabel('y (ln CFU/g) or T (°C)'); % labeling
the x and y axes, respectively
```

More information can be added to the figure for showing, for example, the names of the axes (time, concentration, and temperature). A legend can also be added to identify the symbols and curves. The figure can be shown differently, with a secondary vertical axis (with the *plotyy* command), to differentiate the axes of the microbial concentration and the temperature, for example.

In the second dataset, with the ham temperature decreasing in plateaus from 16 °C to 4 °C, the growth was quantified at times 0, 4.3, 8.5, 11.9, 17.5, 26, 32, 47, 56, 72.9, 104, 144, 168, and

192 h of incubation with the following mean concentration values (duplicate): 8.78, 9.36, 11.09, 12.13, 13.75, 15.64, 16.84, 18.24, 19.15, 19.81, 19.75, 20.27, 19.92, and 20.45 ln CFU/g. In Matlab, for the second dataset, the basic entries of the time (*tdata*) and the microbial concentration (*Cdata*) should be typed in the command window with the following syntax:

```
tdata = [0; 4.3; 8.5; 11.9; 17.5; 26; 32; 47; 56; 72.9; 104;
144; 168; 192]; % times of dataset two
Cdata = [8.78; 9.36; 11.09; 12.13; 13.75; 15.64; 16.84; 18.24;
19.15; 19.81; 19.75; 20.27; 19.92; 20.45]; % concentrations of
dataset two
```

With the same model saved, input information in the Matlab command window should be typed specifying the temperature profile, the time integration range, and the initial conditions of the differential equation for the second dataset.

```
global tp Tp tsize % definition of global variables
tp = [0 11.9 11.9 32.0 32.0 72.9 72.9 192.0]; % time for time-
temperature profile two
Tp = [15.9 15.9 12.9 12.9 8.5 8.5 4.3 4.3]; % temperature for
time-temperature profile two
tsim = [0:0.01:192];
tsize = size(tp);
C0 = Cdata(1);
h0 = 0;
Q0 = log(1+(1/exp(h0)));
ysim = [h0 C0];
[tpred,Cpred] = ode23(@odemodel,tsim,ysim);
```

The results of the growth dynamics simulated for the second dataset can be observed through a figure, similarly to the first case, by typing the following syntax.

```
figure(2) % open a new figure 2 window
plot(tdata, Cdata, 'bo') % plot the symbols of the experimental
dataset, in blue color
hold all % hold all Information in the figure
plot(tpred, Cpred(:,2), 'k-') % plot the continuous line of the
growth predicted by the model, in black color
plot(tp, Tp, 'r--') % plot the dashed line of the time-
temperature profile, in red color
xlabel('t (h)'); ylabel('y (ln CFU/g) or T (°C)'); % labeling
the x and y axes, respectively
```

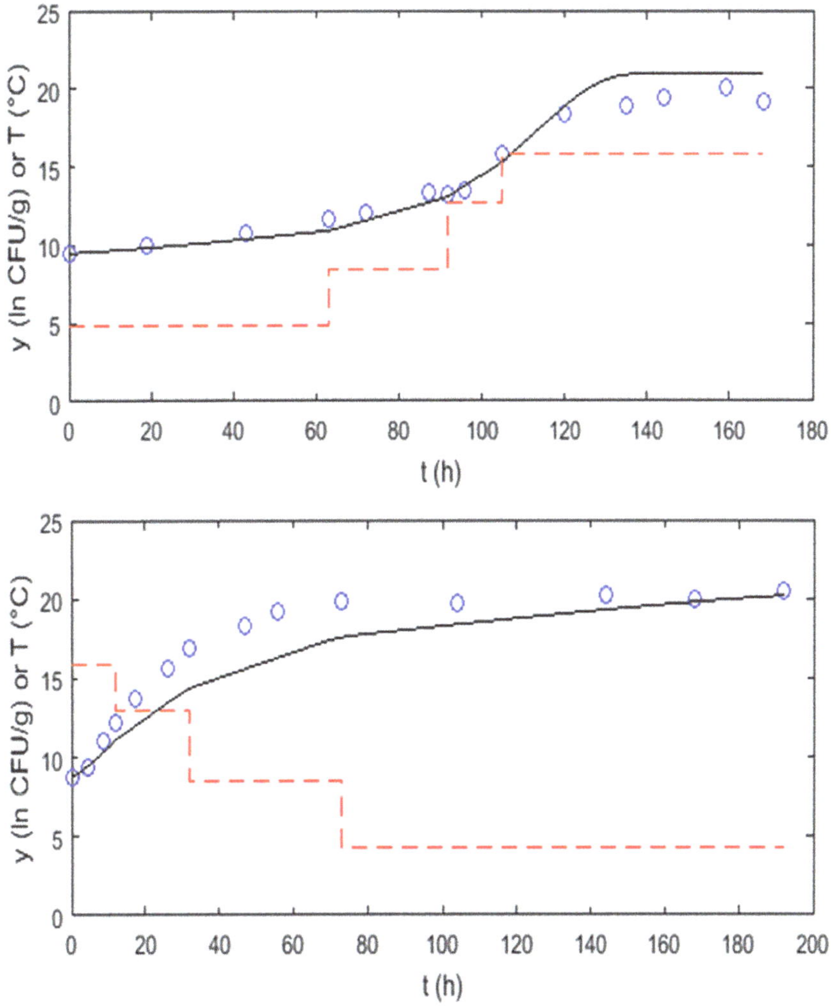

Fig. 4 Two datasets of the growth of *L. viridescens* in vacuum-packed sliced ham (circles) stored at non-isothermal conditions (dashed lines) and the prediction of the Baranyi and Roberts and square-root models (continuous lines) generated in Matlab with the suggested routine. Upper figure: increasing temperature plateaus from 4 °C to 16 °C. Bottom figure: decreasing temperature plateaus from 16 to 4 °C

The figures generated with these procedures in Matlab for the two datasets are shown in Fig. 4. It is important to emphasize that these lines are not fitted curves; they are predictions of the microbial growth obtained under non-isothermal conditions with the primary and secondary models.

4 Notes

1. Many dynamic growth models in the literature were built with the elementary block shown in Eq. 1, which includes parameters of biological meaning (e.g., μ_{max}, λ), but use different mathematical functions (f_{lag} and f_{sta}). Therefore, the value of a given parameter may be different according to the mathematical characteristics of the model.
2. The dynamics of the microbial inactivation can follow a log-linear (shouldered, tailed, shouldered-tailed, biphasic, or biphasic shouldered) or non-log-linear (concave or convex shapes) kinetics, as already mentioned. Therefore, some experiments performed with as much experimental data as possible under constant conditions are desirable, as they contribute to identifying the type/form of the microbial kinetics in the culture medium or food.
3. The applied secondary models depend on how each parameter varies on each factor. For example, the estimates of the y_{max} parameter often show short variation with the temperature; in this case, the arithmetic average of the estimated values may represent the dataset well. The y_0 parameter, in turn, does not need a secondary model since its value can be quantified experimentally, and its result is used as the initial value of the solution of the differential equation of microbial behavior (Eqs. 6 and 8 for growth and inactivation, respectively).
4. Validation of estimated mathematical model parameters with appropriate statistical indices (e.g., bias and accuracy factors, percent bias and percent discrepancy indices) is essential for a reliable prediction of microbial behavior in foods.
5. The mathematical models frequently fail to predict the microbial behavior in foods when there are abrupt variations in the factors, especially the extrinsic ones, such as temperature or gas concentration (see, e.g., [7, 21]). In these cases, additional mathematical functions would be incorporated into the models to adapt the description of microbial behavior in the face of abrupt variations. This topic remains a challenge for predictive microbiology researchers.
6. The initial conditions of the ordinary differential equations of the models interfere with the predictions of microbial behavior in foods. The logarithm of the initial concentration (y_0) of the microorganism is usually quantifiable and is not a big problem, while the initial physiological state of the cells (h_0) is more difficult to quantify/estimate. The h_0 parameter is frequently estimated as an average of the values obtained in previous experiments at constant temperatures.

7. Failures in predicting the microbial behavior in foods can come from different sources, such as the uncertainty associated with experimental data resulting from quantification methods, equipment, and technologies used; the variability of the microbial species associated with the number and characteristics of the strains of each species used in the estimation of the model parameters; and simplifications applied to the mathematical models associated with the factor considerations, such as disregarding temperature heterogeneity in higher foods or disregarding possible changes in factors throughout the experiment (e.g., changes in temperature, pH, or water activity).
8. The parameter estimation with the one-step approach has some advantages, such as (i) parameters can be estimated from experimental datasets under constant and variable conditions; (ii) the estimation is obtained in an objective and optimized way; and (iii) it is possible to minimize the uncertainty about the estimated parameters (confidence intervals) when there are few kinetic datasets available. On the other hand, the main disadvantage is the higher mathematical and numerical complexity of solving the problem.
9. Suppose that temperature data obtained from the datalogger are available. In that case, they can be used as the temperature profile in the Matlab routine instead of considering an equation. For this, they can be saved, for example, in a file named T_profile, in .txt file format, presenting the numerical data of the time in the first column and the temperature data in the second column, without headers or nonnumerical information. Then, in the command window of the Matlab, the text file can be loaded by typing > > *load T_profile.txt*, and then the time data can be loaded with > > *tp = T_profile(:,1)* and temperature with > > *Tp = T_profile(:,2)*.

References

1. Bigelow WD, Esty JR (1920) The thermal death point in relation to time of typical thermophilic organisms. J Infect Dis 27:602
2. Van Impe JF, Nicolai BM, Martens T, De Baerdemaeker J, Vandewalle J (1992) Dynamic mathematical model to predict microbial growth and inactivation during food processing. Appl Environ Microbiol 58:2901
3. Baranyi J, Roberts TA (1994) A dynamic approach to predicting bacterial growth in food. Int J Food Microbiol 23:277
4. Bernaerts K, Dens E, Vereecken K, Geeraerd A, Devlieghere F, Debevere J et al (2003) Modeling microbial dynamics under time-varying conditions. In: Modeling microbial responses in food. Routledge
5. Whiting R, Buchanan R (1993) A classification of models in predictive microbiology – a reply to K. R Davey. Food Microbiol 10(2):175–177
6. Tsoularis A, Wallace J (2002) Analysis of logistic growth models. Math Biosci 179:21
7. Longhi DA, Dalcanton F, Aragão GMFD, Carciofi BAM, Laurindo JB (2013) Assessing the prediction ability of different mathematical models for the growth of *lactobacillus plantarum* under non-isothermal conditions. J Theor Biol 335:88
8. Ranjbaran M, Carciofi BAM, Datta AK (2021) Engineering modeling frameworks for

microbial food safety at various scales. Comprehens Rev Food Sci Food Safety 20(5): 4213–4249. Available from: https://onlinelibrary.wiley.com/doi/10.1111/1541-4337.12818

9. Geeraerd AH, Valdramidis VP, Van Impe JF (2005) GInaFiT, a freeware tool to assess non-log-linear microbial survivor curves. Int J Food Microbiol 102:95
10. Geeraerd AH, Herremans CH, van Impe JF (2000) Structural model requirements to describe microbial inactivation during a mild heat treatment. Intern J Food Microbiol 59(3):185–209. Available from: https://linkinghub.elsevier.com/retrieve/pii/S0168160500003627
11. Baranyi J, Tamplin ML (2004) ComBase: a common database on microbial responses to food environments†. J Food Protect 67(9): 1967–1971. Available from: https://meridian.allenpress.com/jfp/article/67/9/1967/170940/ComBase-A-Common-Database-on-Microbial-Responses
12. Rohatgi A (2021) WebPlotDigitizer [Internet]. Pacifica, California, USA [cited 2022 May 8]. Available from: https://automeris.io/WebPlotDigitizer
13. International Standard Organization (2007) Microbiology of food and animal feeding stuffs - general requirements and guidance for microbiological examinations. ISO 7218
14. International Standard Organization (2006) Microbiology of food, animal feed and water - Preparation, production, storage and performance testing of culture media. ISO 11133
15. International Standard Organization (2019) Microbiology of the food chain – Requirements and guidelines for conducting challange tests of food and feed products – Part 1: Challenge tests to study growth potential, lag time and maximum growth rate. ISO, pp 20976–20971
16. Ratkowsky DA, Olley J, McMeekin TA, Ball A (1982) Relationship between temperature and growth rate of bacterial cultures. J Bacteriol 149:1
17. Rosso L, Lobry JR, Flandrois JP (1993) An unexpected correlation between cardinal temperatures of microbial growth highlighted by a new model. J Theor Biol 162:447
18. Baranyi J, Roberts TA (1995) Mathematics of predictive food microbiology. Intern J Food Microbiol 26(2):199–218. Available from: https://linkinghub.elsevier.com/retrieve/pii/016816059400121L
19. Silva NBD, Longhi DA, Martins WF, Laurindo JB, Aragão GMFD, Carciofi BAM (2017) Modeling the growth of *lactobacillus viridescens* under non-isothermal conditions in vacuum-packed sliced ham. Int J Food Microbiol 240:97
20. Longhi DA, Martins WF, da Silva NB, Carciofi BAM, de Aragão GMF, Laurindo JB (2017) Optimal experimental design for improving the estimation of growth parameters of *lactobacillus viridescens* from data under non-isothermal conditions. Int J Food Microbiol 240:57
21. Zimmermann M, Longhi DA, Schaffner DW, Aragão GMF (2014) Predicting *Bacillus coagulans* spores inactivation in tomato pulp under nonisothermal heat treatments. J Food Sci 79(5):M935

Chapter 9

Mathematical Simulation of the Bio-Protective Effect of Lactic Acid Bacteria on Foodborne Pathogens

Jean Carlos Correia Peres Costa, Araceli Bolívar, and Fernando Pérez-Rodríguez

Abstract

Microbial interaction models are intended to quantify how much the growth of one population is affected by other microbial population(s). In the food context, these models are usually based on two approaches consisting of, on the one hand, the Jameson effect, which represents competition for nutrients between populations (e.g., lactic acid bacteria), and, on the other hand, the Lotka-Volterra (predator-prey) model, in which competition is represented by two inhibition coefficients. Both models and their modifications can be used for predicting microbial interaction under isothermal and non-isothermal conditions in artificial culture media or food. These models can be used as tools to support the assessment and establishment of bio-protective culture-based strategies aimed at reducing the risk of foodborne diseases. Herein, the development and application of microbial interaction models to simulate the bio-protective effect of lactic acid bacteria on foodborne pathogens is described.

Key words Microbial interactions, Lotka-Volterra model, Jameson effect, Predictive microbiology, Bio-preservation, Lactic acid bacteria, Foodborne pathogens

1 Introduction

Food matrices are considered complex microbial ecosystems, where heterogeneous microbial populations coexist and interact with each other and with the environment. In these communities, microorganisms often engage in complex multicellular and intercellular interactions than can affect the behaviour of pathogens in foods [1]. Improving our knowledge on the mechanisms by which microorganisms interact is important to develop more effective food quality and safety management strategies. However, studying microbial interactions and determining theirs exact nature in foods can be challenging and complex, which can lead to significant discrepancies between actual observations and model predictions [2, 3].

Verônica Ortiz Alvarenga (ed.), *Basic Protocols in Predictive Food Microbiology*, Methods and Protocols in Food Science, https://doi.org/10.1007/978-1-0716-3413-4_9,

In foods, microbial interactions can be either nonspecific or specific, which can have special relevance for foodborne pathogens as long as it can determine their increase, survival, or reduction in the product [4, 5]. Nonspecific interaction is associated with physical contact or quorum sensing that induces interactions among microbial communities. It also involves microbial competition for space (i.e., niches) and nutrients. Both phenomena can be affected by changes in initial inoculum concentration (i.e., inoculum size) and environmental conditions. This allows for a selected species to reach stationary phase before others [6]. By contrary, specific interaction does not require physical contact as it is the result from the effect of specific bioactive (antimicrobial) metabolites produced by microbial populations such as lactic acid bacteria (LAB). Species belonging to LAB are able to produce and release bacteriocins and organic acids with the concurrent decrease in pH [7].

Interaction models are usually intended to quantify how much the growth of one population (e.g., foodborne pathogen) is reduced by the growth or presence of other population(s) (e.g., LAB). The models employed to describe microbial interaction are based on two approaches: (i) the Jameson effect, which describes the simultaneous stop of microbial populations when the dominant population reaches the stationary phase considering that one population inhibits the other to the same extent as they inhibit their own growth [8, 9], and (ii) the general Lotka-Volterra competition model (i.e., predator-prey model) for interspecific microbial competition, which includes two empirical parameters, reflecting the degree of interaction or inhibition between both populations [10, 11].

These models can have multiple applications in the food context such as fermentation process optimizations, description of foodborne pathogen and food microbiota interactions, or bio-protective culture formulations.

In the following, the chapter presents the general procedures for developing a microbial interaction model in artificial culture media simulating the bio-protective effect of LAB species against foodborne pathogens.

2 Methods

In previous works, we developed a robust method to evaluate the inhibitory effect of bacteriocin-producing LAB on *L. monocytogenes* using predictive interaction models [12–15]. In brief, to simulate the interaction between LAB species, as bio-protective cultures, and foodborne pathogens, the primary kinetic parameters lag time (λ), maximum specific growth rate (μ_{max}), and maximum population density (N_{max}) are obtained for each microorganism from experimental data in monoculture at different static temperature conditions (*see* Subheading 2.1.2), and using these estimates, secondary models are derived (*see* Subheading 2.2.2). Next, experimental data obtained in coculture are used to estimate competition parameters by means of a regression process (*see* Subheading 2.1.1 and 2.2.4). Finally, both the parameters from the secondary models and the estimated competition parameters can be used to simulate microbial interaction in food under isothermal and non-isothermal conditions. Note that an adjustment factor may be applied to the estimated kinetic parameters to incorporate the food matrix effect.

2.1 Data Generation

The collection of high-quality data is essential for the development of reliable interaction models (*see* **Note 1**). The distribution of collected points within the experimental design should be carefully selected to improve model representativeness while reducing error of the estimated parameters [3]. Data generation is discussed earlier in Chap. 2.

A schematic overview of the experimental design for data generation is presented in Fig. 1.

2.1.1 Media for Microbial Interaction Experiments

Traditionally, microbial interaction studies carried out in model systems are based on the modification of broth culture, which can differ depending on the target microorganisms (e.g., brain heart infusion (BHI) broth or Man, Rogosa, and Sharpe (MRS) broth) and later validated in a specific food matrix. The artificial culture media or food is inoculated with the microorganisms of interest, and the microbial interaction is monitored during storage considering different environmental conditions (e.g., temperatures).

2.1.2 Inoculation and Incubation Conditions to Perform Mono and Coculture Experiments

1. For monoculture experiments, the initial inoculum level of each microorganism should be between ca. 10^2 and 10^3 CFU/mL or g (*see* **Notes 4** and **5**).
2. For coculture experiments, different concentrations of LAB bio-protective culture can be used to evaluate their inhibitory potential against foodborne pathogens. Nevertheless, the initial concentration of the target pathogen is often set lower than the initial concentration of the bio-protective culture (*see* **Note 2**).
3. After inoculation, the studied matrix samples (e.g., artificial culture media or food) should be incubated at different isothermal temperatures, considering representative storage temperatures depending on the studied food (*see* **Note 6**).
4. For experiments in non-isothermal conditions, samples should be incubated with dynamic temperature profiles, considering the studied isothermal temperature range.
5. Storage temperatures should be recorded at regular time intervals using data loggers.

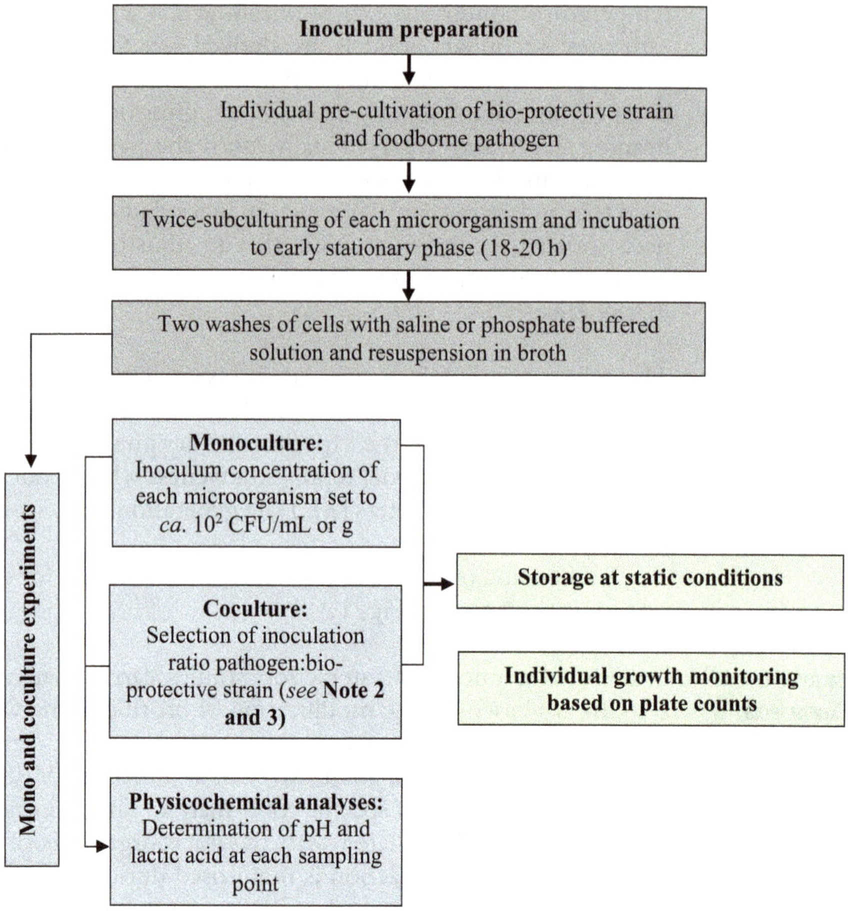

Fig. 1 Schematic illustration of the experimental design for data generation for modelling microbial interaction in foods

2.1.3 Microbiological Analysis in Mono and Coculture Experiments

1. For mono and coculture experiments, at each predetermined sampling point, aliquots must be taken aseptically and serially diluted tenfold in physiological saline water (PSW; 0.85% w/v NaCl) (*see* **Note 7**).
2. Using a pipette, inoculate 0.1 mL of each dilution of the mono- and coculture experiments on the surface of previously prepared Petri dishes, and, as quickly as possible, spread the inoculum over the entire surface of the agar with a Drigalski spatula until the excess liquid is absorbed.
3. Use Man, Rogosa, and Sharpe (MRS) agar at pH 5.7 to enumerate LAB as bio-protective culture and the corresponding selective agar to enumerate the pathogen.
4. Petri dishes should be incubated for approximately 48 h at 30 °C for LAB and corresponding temperature for the target pathogen. Study samples should be sampled periodically until the bio-protective culture reaches the stationary phase.

5. Select the plates of the bio-protective culture and target pathogen with the number of colonies between 25 and 250. Count colonies with the aid of a colony counter magnifying glass, to facilitate the visualization.
6. Convert the plate count values into concentration in logarithmic scale (i.e., log CFU per mL or per g).

2.1.4 Physicochemical Analyses (pH and Lactic Acid)

2.1.4.1 pH Determination

1. Take aliquots of samples from mono- and coculture experiments at regular time intervals to determine pH values using a pH meter. Note that for certain types of samples, a pretreatment or preparation of sample is required before determining pH values.

2.1.4.2 Lactic Acid Determination

1. Take 0.5 mL aliquots from the mono- and coculture samples at regular time intervals, and centrifuge at 10 rpm for 3 min.
2. Dissolve 0.1 mL of the previously centrifuged supernatant in 1.9 ml of iron(III) chloride solution.
3. The amount of lactic acid is measured with a spectrophotometer using the absorbance at 390 nm (*see* **Note 8**).

2.2 Development of Predictive Models

2.2.1 Primary Model for Monoculture Data

1. To estimate the growth parameters for each microorganism (i.e., LAB and pathogen) from monoculture data, fit the Baranyi and Roberts model [16] defined by Eq. 1 to the logarithmically transformed microbial counts (*see* **Note 9**):

$$\begin{cases} \log N_t = \log N_0 + \dfrac{\mu_{max}}{\ln(10)} \cdot F(t) - \dfrac{1}{m \cdot \ln(10)} \\ \quad \cdot \ln\left(1 + \dfrac{e^{m \cdot \mu_{max} \cdot F(t)}}{10^{m(\log N_{max} - \log N_0)}}\right) \\ F(t) = t - \lambda + \dfrac{1}{\mu_{\max}} \cdot \ln\left(1 - e^{-\mu_{\max} \cdot t} + e^{-\mu_{\max} \cdot (t-\lambda)}\right) \end{cases} \quad (1)$$

where N_t is the cellular concentration (CFU/mL or g) at time t, N_0 is the initial concentration (CFU/mL or g), $\mu_{\max}$ is the specific maximum growth rate(h^{-1}), λ is the lag time(h), $N_{\max}$ is the maximum population density (CFU/mL or g), m is a curvature factor, and $F(t)$ represents an adjustment function for the model.

2.2.2 Secondary Models for Monoculture Data

1. To evaluate the influence of storage temperature on the monoculture growth parameters of LAB and pathogen estimated in Subheading 2.2.1, use the square-root model [17] defined by Eq. 2. The model can be fitted by a linear regression using Microsoft Excel® (Microsoft, Redmond, USA):

$$\sqrt{p} = b \cdot (T - T_{\min}) \quad (2)$$

where p is the kinetic parameter (i.e., λ and μ_{max}), b is a constant, T (°C) is temperature, and T_{min} is the theoretical minimum temperature for growth.

2.2.3 Quantification of the Effect of Coculture Conditions on Kinetic Parameters

1. To quantify the growth reduction of pathogen, use the reduction ratio (α), calculated by the complement of the fraction between the parameters obtained in coculture (p_{co}) and monoculture (p_{mono}), as shown by Eq. 3 (*see* **Note 10**):

$$\alpha = 1 - \left(\frac{p_{co}}{p_{mono}}\right) \tag{3}$$

where α is the reduction ratio and p_{co} and p_{mono} are the kinetic parameters (i.e., λ and μ_{max}), in coculture and monoculture, respectively.

2.2.4 Microbial Interaction Models

1. Four existing microbial interaction models, applied in their implicit form, can be used to predict the simultaneous growth of the LAB bio-protective culture and the pathogen, described by the subscript "*sp1*" and "*sp2*," respectively.
2. The Jameson effect model based on Eq. 4 assumes that the pathogen growth stopped when the dominant culture (i.e., LAB) reaches its maximum population density [9]:

$$\begin{cases} \frac{dN_{sp1}}{dt} = N_{sp1} \cdot \mu_{max;sp1} \cdot \left(1 - \frac{N_{sp1}}{N_{max;sp1}}\right) \cdot \left(1 - \frac{N_{sp2}}{N_{max;sp2}}\right) \cdot \left(\frac{Q_{sp1}}{1+Q_{sp1}}\right) \\ \frac{dQ_{sp1}}{dt} = Q_{sp1\,t-1} \cdot \mu_{max;sp1} \\ \frac{dN_{sp2}}{dt} = N_{sp2} \cdot \mu_{max;sp2} \cdot \left(1 - \frac{N_{sp2}}{N_{max:sp2}}\right) \cdot \left(1 - \frac{N_{sp1}}{N_{max;sp1}}\right) \cdot \left(\frac{Q_{sp2}}{1+Q_{sp2}}\right) \\ \frac{dQ_{sp2}}{dt} = Q_{sp2\,t-1} \cdot \mu_{max:sp2} \end{cases} \tag{4}$$

where N is the cell concentration (CFU/mL or g) at time t (h), $\mu_{max;sp}$ is the maximum specific growth rate (h^{-1}), $N_{max;sp}$ is the maximum population density (CFU/mL or g), and Q is a measure of the physiological state of cells at time t (h), for LAB (*sp1*) or pathogen (*sp2*).

The value of Q at $t = 0$ (Q_0) was calculated for both microorganisms as follows:

$$Q_0 = \frac{1}{e^{(\mu_{max} \cdot \lambda)} - 1} \tag{5}$$

3. The modified Jameson effect model described by Eq. 6 with interaction factor incorporates an interaction parameter (γ) to quantify the inhibitory effect of the concentration of LAB on the pathogen growth [18]:

$$\begin{cases} \dfrac{dN_{sp1}}{dt} = N_{sp1} \cdot \mu_{\max;sp1} \cdot \left(1 - \dfrac{N_{sp1}}{N_{\max;sp1}}\right) \cdot \left(1 - \dfrac{\gamma_{12} \cdot N_{sp2}}{N_{\max;sp2}}\right) \cdot \left(\dfrac{Q_{sp1}}{1 + Q_{sp1}}\right) \\ \dfrac{dQ_{sp1}}{dt} = Q_{sp1\ t-1} \cdot \mu_{\max;sp1} \\ \dfrac{dN_{sp2}}{dt} = N_{sp2} \cdot \mu_{\max;sp2} \cdot \left(1 - \dfrac{N_{sp2}}{N_{\max;sp2}}\right) \cdot \left(1 - \dfrac{\gamma_{21} \cdot N_{sp1}}{N_{\max;sp1}}\right) \cdot \left(\dfrac{Q_{sp2}}{1 + Q_{sp2}}\right) \\ \dfrac{dQ_{sp2}}{dt} = Q_{sp2\ t-1} \cdot \mu_{\max;sp2} \end{cases} \tag{6}$$

where γ is the interaction parameter that measures the effect of LAB bio-protective culture on the pathogen (*see* **Note 11**). Other parameters are as indicated for Eq. 4.

4. The modified Jameson effect model with $N_{\text{cri};sp}$, represented by Eq. (7), includes the parameters $N_{\text{cri};sp1}$ and $N_{cri;sp2}$, which refer to the level that LAB ($N_{\text{cri};sp1}$) should achieve to inhibit the pathogen growth and vice versa ($N_{\text{cri};sp2}$) [13, 19]:

$$\begin{cases} \dfrac{dN_{sp1}}{dt} = N_{sp1} \cdot \mu_{\max;sp1} \cdot \left(1 - \dfrac{N_{sp1}}{N_{\max;sp1}}\right) \cdot \left(1 - \dfrac{N_{sp2}}{N_{cri;sp2}}\right) \cdot \left(\dfrac{Q_{sp1}}{1 + Q_{sp1}}\right) \\ \dfrac{dQ_{sp1}}{dt} = Q_{sp1\ t-1} \cdot \mu_{\max;sp1} \\ \dfrac{dN_{sp2}}{dt} = N_{sp2} \cdot \mu_{\max;sp2} \cdot \left(1 - \dfrac{N_{sp2}}{N_{\max;sp2}}\right) \cdot \left(1 - \dfrac{N_{sp1}}{N_{cri;sp1}}\right) \cdot \left(\dfrac{Q_{sp2}}{1 + Q_{sp2}}\right) \\ \dfrac{dQ_{sp2}}{dt} = Q_{sp2\ t-1} \cdot \mu_{\max;sp2} \end{cases} \tag{7}$$

where $N_{\text{cri};sp1}$ and $N_{\text{cri};sp2}$ are the maximum critical concentration (log CFU/mL or g) of LAB on the pathogen and vice versa (*see* **Note 12**). The rest of model parameters are as described for Eq. 4.

5. The Lotka-Volterra model, also known as the predator-prey model, presented in Eq. 8, allows for estimating how the concentration of LAB affects pathogen growth through the addition of a competition factor [20]:

$$\begin{cases} \dfrac{dN_{sp1}}{dt} = N_{sp1} \cdot \mu_{\max;\, sp1} \cdot \left(1 - \dfrac{N_{sp1} + CF_{12} \cdot N_{sp2}}{N_{\max;\, sp1}}\right) \cdot \left(\dfrac{Q_{sp1}}{1 + Q_{sp1}}\right) \\ \dfrac{dQ_{sp1}}{dt} = Q_{sp1\ t-1} \cdot \mu_{\max;\, sp1} \\ \dfrac{dN_{sp2}}{dt} = N_{sp2} \cdot \mu_{\max;\, sp2} \cdot \left(1 - \dfrac{N_{sp2} + CF_{21} \cdot N_{sp1}}{N_{max;\, sp2}}\right) \cdot \left(\dfrac{Q_{sp2}}{1 + Q_{sp2}}\right) \\ \dfrac{dQ_{sp2}}{dt} = Q_{sp2\ t-1} \cdot \mu_{\max;\, sp2} \end{cases} \tag{8}$$

where CF_{12} and CF_{21} are, respectively, the competition factor parameters of LAB on pathogen and vice versa (*see* **Note 13**). The other parameters are as indicated for Eq. 4.

6. The interaction parameters $N_{cri;sp1}$ and $N_{cri;sp2}$ (i.e., maximum critical concentration for each microbial species) and CF_{12} and CF_{21} (i.e., competition factors of one specie on the other) should be estimated by regression but defining primary kinetic parameters (μ_{max}, N_0, N_{max}) with the values obtained from monoculture data (*see* Subheading 2.2.1. *Primary Model Fitting to Monoculture Data* and Subheading 2.2.2. *Secondary Models for Monoculture Data*).
7. The four microbial interaction models are computed in MATLAB using the functions *fmincon* and *ode45* (The MathWorks, Inc.®, Natick, USA).

2.2.5 Prediction Capacity and Model Validation

Many factors, known and unknown, can influence the capacity of predictive microbiology models to accurately describe the microbial response in an actual food. Hence, their adequacy should be previously assessed before application in each specific case. The model validation is usually performed, considering fit-to-purpose mathematical indexes. The main validation indexes widely used in predictive microbiology that can be equally employed to assess the abovementioned interaction models are presented as follows:

1. Root-mean-square error (RMSE), described by Eq. 9, may be used to assess the primary and secondary models:

$$\text{RMSE} = \sqrt{\sum_{i=1}^{n} \frac{\left(y_{\text{obs}} - y_{\text{pred}}\right)^2}{n - p}} \tag{9}$$

where y_{obs} is the observed kinetic value, y_{pred} is the predicted kinetic value, n is the number of observations, and p is the number of parameters of the model.

2. In general, secondary model validation can be also performed by using the statistical indexes bias factor (B_f) and accuracy factor (A_f), defined by Eqs. 10 and 11, respectively (*see* **Note 14**).

$$B_f = 10^{\left(\sum_{i=1}^{n} \log\left(\frac{y_{\text{pred}}}{y_{\text{obs}}}\right)/n\right)} \tag{10}$$

$$A_f = 10^{\left(\sum_{i=1}^{n} \left|\log\left(\frac{y_{\text{pred}}}{y_{\text{obs}}}\right)\right|/n\right)} \tag{11}$$

3. The parameters of these equations were defined for Eq. 9. Interaction models tested under isothermal temperature conditions can be assessed with RMSE (Eq. 9) and the corrected Akaike information criteria (AICc) defined by Eq. 12:

$$\text{AIC}_c = (n)\ln\left(\frac{\text{SSE}}{n}\right) + 2(p+1) + \frac{2(p+1)\cdot(p+2)}{n-p-2} \tag{12}$$

where SSE is the sum of squares, and the other parameters of this equation are as indicated for Eq. 9.

4. The acceptable simulation zone (ASZ) approach is rather suitable for validation of the prediction capacity of the interaction models at non-isothermal conditions (*see* **Note 15**).

3 Application

In the European Union, the Regulation (EC) 2073/2005 (amended by Regulation (EC) 1441/2007) lays food safety criteria for foodstuffs, whereby food business operators (FBOs) are responsible for demonstrating that foodborne pathogens should not be present or that do not exceed established levels during the shelf life of the food. In this context, the use of LAB as bio-protective cultures assisted by the use of interaction models is proposed as a promising strategy to ensure food safety of different RTE foods. A summary of studies modelling the microbial interaction between LAB, as bio-protective cultures, and pathogens in artificial culture medium and food is shown in Table 1.

4 Notes

1. For the development of kinetics models, it is recommended that, for a particular combination of conditions, a minimum of ten points should be collected.
2. For coculture experiments, it is recommended that the initial pathogen concentration be ca. 10^2 or 10^3 CFU/mL, while for the LAB, three different concentrations can be investigated

Table 1
Microbial interaction and inhibition described by using predictive interaction models of LAB as bio-protective cultures and foodborne pathogens in artificial culture media and different food matrices

Matrix	Lactic acid bacteria (LAB)	Foodborne pathogen	Type model	Reference
Pork meat products[1]	Lactic acid microbiota	*Listeria monocytogenes*	Variants of the Jameson effect and Lotka-Volterra models	[20]
Lightly preserved seafood	LAB cocktail[2]		Simple Jameson effect model	[21]
Frankfurter sausages	*Leuconostoc carnosum*		Dens et al. model based on Lotka-Volterra model	[22]
Minas fresh cheese	LAB cocktail[3]		Modified Jameson effect model with γ factor	[23]
Vacuum packed hot-smoked sea bream	*Latilactobacillus sakei* CTC494		Modified Jameson effect model with N_{cri} and simple Lotka-Volterra model	[14]
Vacuum-packaged cooked ham	*L. sakei* CTC494	*L. monocytogenes*	Modified Jameson effect model with γ factor	[24]
Cold-smoked salmon	*Lactobacillus* spp. and *Carnobacterium* spp.		Simple Jameson effect	[9]
Fresh pork	Natural microbiota	*Salmonella* spp.	Simple Jameson effect, modified Jameson effect model with γ factor, and Classical Lotka-Volterra model	[18]
Milk	Starter culture of lactic acid bacteria[4]	*Staphylococcus aureus*	Modified Jameson effect model with *CPD*[5]	[19]
Culture media broth	*L. sakei* 115	*L. monocytogenes*	Modified Jameson effect model with N_{cri} and simple Lotka-Volterra model	[13]
Cottage cheese	LAB cocktail[6]		Simple Jameson effect	[25]
Model meat gravy	*L. sakei* MN		Modified Jameson effect model with β factor	[26]
Fish juice[7]	*L. sakei* CTC494		Classical Lotka-Volterra model	[12]
Surimi and tuna pâté	*L. sakei* CTC494		Modified Jameson effect model with γ factor and modified of adjustment function including an empirical parameter β	[15]

(continued)

Table 1 (continued)

Matrix	Lactic acid bacteria (LAB)	Foodborne pathogen	Type model	Reference
Vacuum-packaged chilled pork	*Lactobacillus* spp.	*L. monocytogenes*	Modified Jameson effect model and modified Lotka-Volterra model	[27]
Artisanal Minas semihard cheese	LAB cocktail[3]		Modified Jameson effect model with γ factor	[28]
Processed seafood and mayonnaise-based seafood salads	Psychrotolerant lactic acid bacteria[8]		Modified Jameson effect model with γ factor	[29]
Modified culture media broth	*L. sakei* and *Staphylococcus carnosus*		Modified Jameson effect model with interspecific competition term, β	[30]
Minced tuna	Natural microbiota		Modified Jameson effect model	[31]
Gilthead seabream	Aerobic natural microbiota	*Aeromonas hydrophila*	Classical Lotka-Volterra model	[32]
Milk	LAB cocktail[9]	*S. aureus and Escherichia coli*	Simple and modified Jameson effect model with competition coefficients	[33]
Minced pork	*Leuconostoc gelidum*	*Brochothrix thermosphacta* and *Pseudomonas* spp.	Modified Jameson effect model with N_{cri} and classical Lotka-Volterra model	[34]
Milk	LAB cocktail[9]	*E. coli*	Simple and modified Jameson effect model with competition coefficients	[35]

[1]Pork meat product included cooked and smoked diced bacon and salted diced bacon.

[2]In this study lactic acid bacteria included *Latilactobacillus sakei*, *Latilactobacillus curvatus*, *Carnobacterium maltaromaticum*, *Enterococcus malodoratus* and *Leuconostoc* spp.

[3]Cultures of lactic acid bacteria in Minas fresh cheese included *Latilactobacillus brevis*, *Lactiplantibacillus plantarum* and *Enterococcus faecalis.*

[4]Starter culture consists of *Lactococcus lactis* subsp. *lactis* and *Streptococcus salivarius* subsp. *thermophilus.*

[5]CPD refers to critical population density..

[6]Cultures of lactic acid bacteria in cottage cheese included *L. lactis* subsp. *lactis*, *L. lactis* subsp. *cremoris* and *L. lactis* subsp. *lactis* biovar. *diacetylactis.*

[7]Study carried out on fish juice and validated on gilthead sea bream (*Sparus aurata*).

[8]Psychrotolerant lactic acid bacteria included four isolates of *Latilactobacillus sakei* (LA-5, A1, MAP23-1 and F46-4).

[9]Cultures of lactic acid bacteria in milk included *L. lactis* subsp. *lactis, L. lactis* subsp. *cremosis* and *Streptococcus thermophilus.*

(e.g., 10^2, 10^4, and 10^6 CFU/mL), thus generating three (initial) inoculation ratios of pathogen/bio-protective culture, corresponding to 1:1, 1:2, and 1:3 in logarithmic scale.

3. Importantly, before applying LAB as bio-protective cultures, it is crucial to assess their sensory impact on the target food product.
4. It is recommended that the microbial inoculum volume is as low as possible, i.e., $\leq 1\%$ (v/v or w/v).
5. If the inoculum level is too low, microorganisms could not present growth under the conditions tested, due to the increase in the lag time which can be affected by product formulation [3].
6. From an application standpoint, lactic acid bacteria species with better growth ability at low temperatures should be selected for bio-preservation of refrigerated foods.
7. In microbial interaction studies, plate counting is the golden standard method to monitor the simultaneous growth of microbial populations, based on the use of selective media for differentiating the target microorganisms.
8. The method for determining lactic acid is based on the spectrophotometric determination of the colored product of the reaction of lactate ions and iron(III) chloride at 390 nm. To obtain optimal reaction conditions, it is necessary to construct a calibration curve (*see* [36]).
9. The kinetic parameters based on the Baranyi model of mono- and coculture experiments can be obtained using the DMFit Excel add-in, available at http://www.combase.cc.
10. The kinetic parameters of coculture experiments can also be obtained by the Baranyi model (*see* Subheading 2.2.1. *Primary Model Fitting to Monoculture Data*).
11. The modified Jameson effect model with interaction γ factor (Eq. 6) is more flexible than the original Jameson effect model since it includes an interaction parameter (γ) that has to be estimated from microbial growth curves in coculture. This parameter allows describing pathogen concentration increase ($\gamma < 1$) or decrease ($\gamma > 1$) after the LAB bio-protective culture has reached its $N_{\max}$. With $\gamma = 1$, both models are equivalent, i.e., the pathogen concentration stabilizes at the actual maximum population density value after LAB bio-protective culture has reached its N_{max}.
12. The modified Jameson effect model (Eq. 7) replaces $N_{max;sp1}$ and $N_{max;sp2}$ by the parameters $N_{cri;sp1}$ and $N_{cri;sp2}$, describing the maximum critical concentration that a microbial population should reach to inhibit the growth of the other microbial population. The $N_{cri;sp1}$ and $N_{cri;sp2}$ parameters are usually lower than $N_{max;sp1}$ and $N_{max;sp2}$, respectively.

13. The Lotka-Volterra model describes through the competition factor (CF_{12}) different interaction responses. The pathogen can stop growing ($CF_{12} = 1$), grow with reduced μ_{max} ($0 < CF_{12} < 1$), or decline its population when LAB reaches its N_{max} ($CF_{12} > 1$).
14. The bias factor (B_f) indicates whether the mathematical model tends to overestimate (safe failure, $B_f > 1$) or underestimate growth (dangerous failure, $B_f < 1$); and the accuracy factor (A_f), is a measure of the average difference between the observed and predicted values. For instance, an A_f of 1.5 indicates that the prediction is, on average, a factor of 1.5 different from the observed value. Both indices are valuable and aim to evaluate the performance of the fitted models [37].
15. Model performance is considered acceptable when a least 70% of the observed log count values are within the ASZ, defined as ±0.5 log units from the predicted concentration in log units [18].

Acknowledgments

This work was financed by the project AT21-00227 (BioProtect) and the project P18-RT-1386 (RESISTALI), both supported by, Consejería de Economía, Conocimiento, Empresas y Universidad (Andalusia, Spain) and the European Regional Development Fund (ERDF). Author **J.C.C.P. Costa** was supported by the Ministry of Universities (Government of Spain) through the *Plan de Recuperación, Transformación y Resiliencia* funded by the European Union – NexGenerationEU.

References

1. Zuñiga C, Zaramela L, Zengler K (2017) Elucidation of complexity and prediction of interactions in microbial communities. Microb Biotechnol 10:1500. https://doi.org/10.1111/1751-7915.12855
2. Powell M, Schlosser W, Ebel E (2004) Considering the complexity of microbial community dynamics in food safety risk assessment. Int J Food Microbiol 90:171. https://doi.org/10.1016/S0168-1605(03)00106-5
3. Pérez-Rodríguez F, Valero A (2013) Predictive microbiology in foods. Springer Briefs in Food, Health, and Nutrition Series. Springer (New York) 978-1-4614-5520-2. https://doi.org/10.1007/978-1-4614-5520-2
4. Haruta S, Kato S, Yamamoto K, Igarashi Y (2009) Intertwined interspecies relationships: Approaches to untangle the microbial network. Environ Microbiol 11:2963. https://doi.org/10.1111/j.1462-2920.2009.01956.x
5. Zilelidou EA, Skandamis PN (2018) Growth, detection and virulence of *Listeria monocytogenes* in the presence of other microorganisms: microbial interactions from species to strain level. Int J Food Microbiol 277:10. https://doi.org/10.1016/j.ijfoodmicro.2018.04.011
6. Mellefont LA, McMeekin TA, Ross T (2008) Effect of relative inoculum concentration on *Listeria monocytogenes* growth in co-culture. Int J Food Microbiol 121:157. https://doi.org/10.1016/j.ijfoodmicro.2007.10.010
7. Kasra-Kermanshahi R, Mobarak-Qamsari E (2015) Inhibition effect of Lactic acid bacteria against food born pathogen, *Listeria monocytogenes*. Appl Food Biotechnol 2:11. https://doi.org/10.22037/afb.v2i4.8894

8. Jameson JE (1962) A discussion of the dynamics of salmonella enrichment. J Hyg (Lond) 60: 193. https://doi.org/10.1017/S0022172400039462
9. Giménez B, Dalgaard P (2004) Modelling and predicting the simultaneous growth of *Listeria monocytogenes* and spoilage micro-organisms in cold-smoked salmon. J Appl Microbiol 96:96. https://doi.org/10.1046/j.1365-2672.2003.02137.x
10. Dens EJ, Vereecken KM, Van Impe JF (1999) A prototype model structure for mixed microbial populations in homogeneous food products. J Theor Biol 201:159. https://doi.org/10.1006/jtbi.1999.1021
11. Vereecken KM, Dens EJ, Van Impe JF (2000) Predictive modeling of mixed microbial populations in food products: Evaluation of two-species models. J Theor Biol 205:53. https://doi.org/10.1006/jtbi.2000.2046
12. Costa JCCP, Bover-Cid S, Bolívar A et al (2019) Modelling the interaction of the sakacin-producing *Lactobacillus sakei* CTC494 and *Listeria monocytogenes* in filleted gilthead sea bream (Sparus aurata) under modified atmosphere packaging at isothermal and non-isothermal conditions. Int J Food Microbiol 297:72. https://doi.org/10.1016/j.ijfoodmicro.2019.03.002
13. Costa JCCP, Bolívar A, Valero A et al (2020) Evaluation of the effect of *Lactobacillus sakei* strain L115 on *Listeria monocytogenes* at different conditions of temperature by using predictive interaction models. Food Res Int 131: 108928. https://doi.org/10.1016/j.foodres.2019.108928
14. Bolívar A, Costa JCCP, Posada-Izquierdo GD et al (2021) Quantifying the bioprotective effect of *Lactobacillus sakei* CTC494 against *Listeria monocytogenes* on vacuum packaged hot-smoked sea bream. Food Microbiol 94: 103649. https://doi.org/10.1016/j.fm.2020.103649
15. Bolívar A, Tarlak F, Costa JCCP et al (2021) A new expanded modelling approach for investigating the bioprotective capacity of *Latilactobacillus sakei* CTC494 against *Listeria monocytogenes* in ready-to-eat fish products. Food Res Int 147:110545. https://doi.org/10.1016/j.foodres.2021.110545
16. Baranyi J, Roberts TA (1994) A dynamic approach to predicting bacterial growth in food. Int J Food Microbiol 23:277. https://doi.org/10.1016/0168-1605(94)90157-0
17. Ratkowsky DA, Olley J, McMeekin TA, Ball A (1982) Relationship between temperature and growth rate of bacterial cultures. J Bacteriol 149:1. https://doi.org/10.1128/jb.149.1.1-5.1982
18. Møller COA, Ilg Y, Aabo S et al (2013) Effect of natural microbiota on growth of *Salmonella* spp. in fresh pork – A predictive microbiology approach. Food Microbiol. https://doi.org/10.1016/j.fm.2012.10.010
19. Le Marc Y, Valík L, Medveďová A (2009) Modelling the effect of the starter culture on the growth of *Staphylococcus aureus* in milk. Int J Food Microbiol 129:306. https://doi.org/10.1016/j.ijfoodmicro.2008.12.015
20. Cornu M, Billoir E, Bergis H et al (2011) Modeling microbial competition in food: Application to the behavior of *Listeria monocytogenes* and lactic acid flora in pork meat products. Food Microbiol 28:639–647
21. Mejlholm O, Dalgaard P (2007) Modeling and predicting the growth of lactic acid bacteria in lightly preserved seafood and their inhibiting effect on *Listeria monocytogenes.* J Food Prot 70:2485. https://doi.org/10.4315/0362-028X-70.11.2485
22. Baka M, Noriega E, Mertens L et al (2014) Protective role of indigenous *Leuconostoc carnosum* against *Listeria monocytogenes* on vacuum packed Frankfurter sausages at suboptimal temperatures. Food Res Int 66: 197. https://doi.org/10.1016/j.foodres.2014.08.011
23. Cadavez VAP, Campagnollo FB, Silva RA et al (2019) A comparison of dynamic tertiary and competition models for describing the fate of *Listeria monocytogenes* in Minas fresh cheese during refrigerated storage. Food Microbiol 79:48. https://doi.org/10.1016/j.fm.2018.11.004
24. Serra-Castelló C, Costa JCCP, Jofré A et al (2022) A mathematical model to predict the antilisteria bioprotective effect of *Latilactobacillus sakei* CTC494 in vacuum packaged cooked ham. Int J Food Microbiol 363: 109491. https://doi.org/10.1016/j.ijfoodmicro.2021.109491
25. Østergaard NB, Eklöw A, Dalgaard P (2014) Modelling the effect of lactic acid bacteria from starter- and aroma culture on growth of *Listeria monocytogenes* in cottage cheese. Int J Food Microbiol 188:15. https://doi.org/10.1016/j.ijfoodmicro.2014.07.012
26. Quinto EJ, Marín JM, Schaffner DW (2016) Effect of the competitive growth of *Lactobacillus sakei* MN on the growth kinetics of *Listeria monocytogenes* Scott A in model meat gravy. Food Control 63:34. https://doi.org/10.1016/j.foodcont.2015.11.025
27. Ye K, Wang H, Jiang Y et al (2014) Development of interspecific competition models for the growth of *Listeria monocytogenes* and *Lactobacillus* on vacuum-packaged chilled pork by quantitative real-time PCR. Food Res Int 64:

626. https://doi.org/10.1016/j.foodres.2014.07.017

28. Gonzales-Barron U, Campagnollo FB, Schaffner DW et al (2020) Behavior of *Listeria monocytogenes* in the presence or not of intentionally-added lactic acid bacteria during ripening of artisanal Minas semi-hard cheese. Food Microbiol 91:103545. https://doi.org/10.1016/j.fm.2020.103545
29. Mejlholm O, Dalgaard P (2015) Modelling and predicting the simultaneous growth of *Listeria monocytogenes* and psychrotolerant lactic acid bacteria in processed seafood and mayonnaise-based seafood salads. Food Microbiol 46:1. https://doi.org/10.1016/j.fm.2014.07.005
30. Blanco-Lizarazo CM, Sotelo-Díaz I, Llorente-Bousquets A (2016) In vitro modelling of simultaneous interactions of *Listeria monocytogenes*, *Lactobacillus sakei*, and *Staphylococcus carnosus*. Food Sci Biotechnol 25:341. https://doi.org/10.1007/s10068-016-0048-0
31. Koseki S, Takizawa Y, Miya S et al (2011) Modeling and predicting the simultaneous growth of *Listeria monocytogenes* and natural flora in minced tuna. J Food Prot 74:176. https://doi.org/10.4315/0362-028X.JFP-10-258
32. Giuffrida A, Ziino G, Valenti D et al (2007) Application of an interspecific competition model to predict the growth of *Aeromonas hydrophila* on fish surfaces during refrigerated storage. Arch Leb. https://doi.org/10.2377/0003-925X-58-136
33. Ačai P, Medved'ová A, Mančušková T, Valík L (2019) Growth prediction of two bacterial populations in co-culture with lactic acid bacteria. Food Sci Technol Int 25:692. https://doi.org/10.1177/1082013219860360
34. Cauchie E, Delhalle L, Baré G et al (2020) Modeling the Growth and Interaction Between *Brochothrix thermosphacta*, *Pseudomonas* spp., and *Leuconostoc gelidum* in Minced Pork Samples. Front Microbiol. https://doi.org/10.3389/fmicb.2020.00639
35. Ačai P, Valík L, Medved'Ová A, Rosskopf F (2016) Modelling and predicting the simultaneous growth of *Escherichia coli* and lactic acid bacteria in milk. Food Sci Technol Int 22:475. https://doi.org/10.1177/1082013215622840
36. Borshchevskaya LN, Gordeeva TL, Kalinina AN, Sineokii SP (2016) Spectrophotometric determination of lactic acid. J Anal Chem 71:755. https://doi.org/10.1134/S1061934816080037
37. Haberbeck LU, Plaza-Rodríguez C, Desvignes V et al (2018) Harmonized terms, concepts and metadata for microbiological risk assessment models: The basis for knowledge integration and exchange. Microb Risk Anal 10:3. https://doi.org/10.1016/j.mran.2018.06.001

Chapter 10

Acceptable Prediction Zones Method for the Validation of Predictive Models for Foodborne Pathogens

Thomas P. Oscar

Abstract

Proper validation of models is important because it provides users with confidence that model predictions are reliable. In addition, it helps modelers identify and repair prediction problems before models are provided to end users. This chapter describes the acceptable prediction zones (APZ) method in the validation software tool (ValT) for predictive microbiology. The APZ method is the first validation method to have criteria for test data, model performance, and model validation. These criteria ensure that comparisons of observed and predicted values are not confounded by differences in data collection and modeling methods. In addition, they ensure that model validation is accurate, unbiased, and objective. A tertiary model and secondary models (lag time, growth rate) for growth of *Salmonella* Typhimurium definitive phage type 104 (DT104) on chicken skin with native microflora are used to demonstrate the APZ method in ValT.

Key words Acceptable prediction zones method, Predictive model, Model validation, Validation software tool

1 Introduction

Proper validation of predictive models for foodborne pathogens is important for two reasons. First, it provides model developers with an objective method to identify and repair prediction problems before models are provided to end users [1–3]. Second, it provides end users with confidence that model predictions are reliable [4, 5].

There are three classes of predictive models for foodborne pathogens: (1) primary, (2) secondary, and (3) tertiary [6]. Primary models predict log number as a function of time and one combination of independent variables (temperature, pH, water activity, gas atmosphere) [7, 8]. Secondary models predict primary model parameters (lag time, growth rate) as a function of independent variables [9–12]. Tertiary models predict log number as a function of time and all combinations of independent variables and are developed in user-friendly software applications by incorporating

Verônica Ortiz Alvarenga (ed.), *Basic Protocols in Predictive Food Microbiology*, Methods and Protocols in Food Science, https://doi.org/10.1007/978-1-0716-3413-4_10,

secondary models back into the primary model from which they were developed [13, 14].

Validation of models for foodborne pathogens involves comparison of observed and predicted values [15–17]. The residual (observed – predicted) is used to compare observed and predicted values from primary, secondary (initial number, maximum population density), and tertiary models than predict log number. On the other hand, the relative residual is used to compare observed and predicted values from secondary models that predict lag time ((predicted – observed)/predicted) or growth rate ((observed – predicted)/predicted).

Model validation occurs in three steps with three types of data: (1) dependent data, (2) independent data for interpolation, and (3) independent data for extrapolation [18]. The first two steps are required for model validation. Although the third step is optional, it is important because it can save time and money by finding independent variables for which new models are not needed. For example, if a model developed with strain A was validated for extrapolation to strain B, a new model would not be needed for strain B.

Model validation involves three types of criteria: (1) test data, (2) model performance, and (3) model validation [18]. Criteria for test data ensure that comparisons of observed and predicted values are not confounded by differences in data collection and modeling methods and that evaluation of model performance is accurate and unbiased. Criteria for model performance ensure an objective evaluation of model performance, whereas criteria for model validation ensure an objective decision about model validation.

The acceptable prediction zones (APZ) method is the first validation method to have criteria for test data, model performance, and model validation [18, 19]. Use of the APZ method is growing [20–23] but not all criteria are being used. To facilitate proper use of the APZ method, the validation software tool or ValT (vault) for predictive microbiology was developed [24]. The focus of this chapter is to describe and demonstrate how the APZ method in ValT can be used to guide development and validation of secondary models for lag time and growth rate and tertiary models for log number.

2 Materials

There are three versions of ValT: (1) LN for primary, secondary, and tertiary models; (2) LT for secondary models for lag time; and (3) GR for secondary models for growth rate. Once this chapter is published, copies of all versions of ValT will be available at no cost at www.ars.usda.gov/nea/errc/PoultryFARM. Excel 2016 or higher (Microsoft Corp., Redmond, WA) is needed to run ValT.

3 Method

3.1 Terms

1. *Dependent data* are data used in model development.
2. *Independent data* are data not used in model development.
3. *Interpolation data* are independent data that are within the ranges of independent variables used in model development.
4. *Extrapolation data* are independent data that are outside the ranges of independent variables used in model development.
5. *Prediction case* is a pair of observed and predicted values.
6. *Special prediction case* is a prediction case for which a prediction error cannot be calculated (*see* **Note 1**).
7. *Prediction error* is a metric used to compare observed and predicted values.
8. *Residual* is the prediction error used for models that predict log number (*see* **Note 1**).
9. *Relative residual* is the prediction error used for models that predict lag time or growth rate (*see* **Note 2**).
10. *Acceptable prediction zones* (APZ) is a range of prediction errors that are fully or partly acceptable (*see* **Note 3**).
11. *APZ value* is a number from 0 to 1 that is assigned to a prediction case based on the location of its prediction error in or out of the APZ (*see* **Note 4**).
12. pAPZ is the proportion of prediction errors in the APZ or the average APZ value for a set of test data.
13. *Fail-safe prediction* is a model prediction that underpredicts food safety.
14. *Fail-dangerous prediction* is a model prediction that overpredicts food safety.

3.2 Validation Software Tool (ValT)

It is a user-friendly, computer software tool to facilitate proper use of the APZ method. It was developed in an Excel notebook and consists of eight spreadsheets. There are three versions of ValT, one for primary, secondary, and tertiary models that predict log number ($ValT_{LN}$), one for secondary models that predict lag time ($ValT_{LT}$), and one for secondary models that predict growth rate ($ValT_{GR}$). All three versions of ValT are similar.

1. Spreadsheet 1 (Fig. 1) is the Table of Contents with a graph and description of the APZ method.
2. Spreadsheet 2 (Fig. 2) is for input of dependent data and calculation of prediction errors and APZ values. This includes entry of values for independent variables and corresponding observed and predicted values.

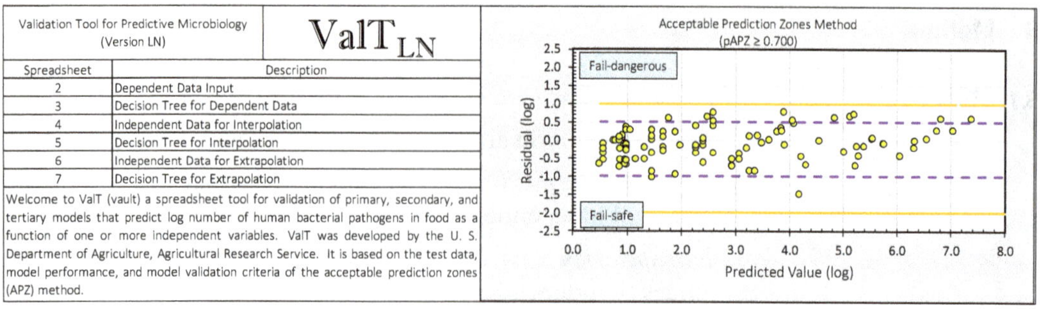

Fig. 1 Screenshot of the Table of Contents spreadsheet in the validation software tool (ValT) for primary, secondary, and tertiary models that predict log number. This spreadsheet is similar to the other two versions of ValT for secondary models for lag time and growth rate except the figure is modified to reflect differences in prediction errors and boundaries of the acceptable prediction zones (APZ)

Temp	Time	Observed	Predicted	Residual	APZ
5	0	0.94	0.85	0.09	1.00
5	0	0.90	0.85	0.05	1.00
5	2	0.55	0.85	-0.30	1.00
5	2	0.98	0.85	0.13	1.00
5	4	1.66	0.85	0.81	0.39
5	4	0.98	0.85	0.13	1.00
5	6	1.66	0.85	0.81	0.39
5	6	0.55	0.85	-0.30	1.00
5	8	0.98	0.85	0.13	1.00
5	8	0.98	0.85	0.13	1.00
5	0	0.85	0.85	0.00	1.00

Fig. 2 Screenshot of the spreadsheet for input of dependent data in the validation software tool (ValT) for primary, secondary, and tertiary models that predict log number. Data for values of independent variables (temperature, time) and observed and predicted log number are entered, and then the spreadsheet calculates the prediction error (residual) and assigns an acceptable prediction zones (APZ) value from 0 to 1. This spreadsheet is similar to other types of data (independent data for interpolation, independent data for extrapolation) and for other types of models (secondary models for lag time and growth rate) except the prediction error (relative residual) is different for the other types of models

3. Spreadsheet 3 (Fig. 3) has a decision tree, a pivot table, and instructions (*see* **Note 5**) for evaluating dependent data.
   (a) The decision tree contains a series of questions for test data, model performance, and model validation criteria.
   (b) The pivot table is used to count prediction cases per combination and level of independent variables and

Q	A	Decision Tree for Dependent Data
1	yes	Were the data used to develop the model?
2	yes	Were the independent variables evenly spaced?
3	no	Was there a minimum of four prediction cases per combination of independent variables?
4	no	Did all combinations of independent variables have the same number of prediction cases?
5	yes	Was the overall pAPZ ≥ 0.70?
6	yes	Was pAPZ for all individual levels of independent variables ≥ 0.70?
7	no	Was a single pAPZ ≥ 0.70 for every three consecutive combinations of the independent variables?
8	no	Was the model validated for dependent data?

Average of APZ	Time					
Temp	0	2	4	6	8	Average
5	1.00	0.95	0.85	0.85	1.00	**0.92**
10	1.00	0.98	0.82	0.97	0.88	**0.93**
15	1.00	0.82	0.97	0.95	0.95	**0.94**
20	1.00	0.97	0.85	0.90	0.49	**0.84**
25	1.00	0.97	1.00	0.80	0.97	**0.95**
30	1.00	1.00	0.97	0.89	0.93	**0.96**
35	1.00	1.00	1.00	1.00	0.94	**0.99**
40	1.00	1.00	0.99	0.84	0.97	**0.96**
45	1.00	0.95	0.99	0.84	0.73	**0.91**
50	1.00	0.85	0.57	0.67	0.65	**0.75**
Average	**1.00**	**0.95**	**0.91**	**0.87**	**0.85**	**0.91**

Fig. 3 Screenshot of the spreadsheet in the validation software tool (ValT) that has the decision tree and pivot table for primary, secondary, and tertiary models that predict log number. The decision tree contains "yes" or "no" questions (Q) for test data criteria (Q1 to Q4), for model performance criteria (Q5 to Q7), and for model validation criteria (Q8). The pivot table shows results for average acceptable prediction zones (APZ) values for individual combinations of independent variables (temperature, time), for individual levels of independent variables, and overall. This spreadsheet is similar in other versions of ValT for secondary models that predict lag time or growth rate

overall. These results are used to answer questions for test data criteria (Q1 to Q4). In addition, the pivot table is used to calculate pAPZ for each combination and level of independent variables and overall. These results are used to answer questions for model performance criteria (Q5 to Q7).

(c) Each question for test data and model performance criteria has a pull-down menu for selecting a "yes" or "no" answer. The "yes" or "no" question (Q8) for model validation is answered by ValT using a formula.

(d) A model is validated for dependent data when answers to all questions in the decision tree are "yes."

4. Spreadsheet 4 (not shown) is where independent data for interpolation are entered. The rest of the description is the same as for spreadsheet 2.

5. Spreadsheet 5 (Fig. 4) has a decision tree, a pivot table, and instructions (*see* **Note 6**) for evaluating independent data for interpolation. The rest of the description is the same as for spreadsheet 3 except that questions for test data criteria are

Q	A	Decision Tree for Interpolation
1	no	Was the model validated for dependent data?
2	yes	Were the data independent?
3	yes	Were the data collected using the same methods as dependent data?
4	no	Were the independent variables at values intermediate to those used in model development?
5	yes	Was there a minimum of two prediction cases per combination of independent variables?
6	no	Did all combinations of independent variables have the same number of prediction cases?
7	yes	Was the overall pAPZ ≥ 0.70?
8	yes	Was pAPZ for all individual levels of independent variables ≥ 0.70?
9	yes	Was a single pAPZ ≥ 0.70 for every three consecutive combinations of the independent variables?
10	no	Was the model validated for interpolation?

Average of APZ	Time					
Temp	0	2	4	6	8	Average
7.5	1.00	1.00	0.95	1.00	0.95	0.98
12.5	1.00	0.84	0.90	0.95	1.00	0.94
17.5	1.00	0.95	1.00	0.90	0.80	0.93
22.5	1.00	1.00	0.98	1.00	0.89	0.97
27.5	1.00	0.98	1.00	1.00	0.46	0.89
32.5	1.00	0.98	1.00	0.98	0.97	0.99
37.5	1.00	1.00	1.00	0.93	0.97	0.98
42.5	1.00	0.88	1.00	0.81	0.72	0.89
47.5	1.00	1.00	1.00	0.82	0.56	0.88
Average	1.00	0.96	0.98	0.93	0.81	0.94

Fig. 4 Screenshot of the spreadsheet in the validation software tool (ValT) that has a decision tree and pivot table for independent data for interpolation for primary, secondary, and tertiary models that predict log number. The decision tree contains "yes" or "no" questions (Q) for test data criteria (Q2 to Q6), model performance criteria (Q7 to Q9), and model validation criteria (Q1 and Q10). The pivot table shows results for average acceptable prediction zones (APZ) values for individual combinations of independent variables (temperature, time), for individual levels of independent variables, and overall. This spreadsheet is similar in other versions of ValT for secondary models that predict lag time or growth rate

Q2 to Q6, questions for model performance criteria are Q7 to Q9, and questions for model validation criteria are Q1 and Q10.

6. Spreadsheet 6 (not shown) is where independent data for extrapolation are entered. The rest of the description is the same as for spreadsheet 2.
7. Spreadsheet 7 (Fig. 5) has a decision tree, a pivot table, and instructions (*see* **Note 6**) for evaluating independent data for extrapolation. The rest of the description is the same as for spreadsheet 5.
8. Spreadsheet 8 (not shown) contains an array of prediction errors (residual or relative residual) and corresponding APZ values.
   (a) Spreadsheets 2, 4, and 6 contain a formula (*see* **Note 7**) for assigning APZ values including for special prediction cases. Those formula are linked to spreadsheet 8!
   (b) The array and formula for assigning APZ values differ slightly for each type of model (log number, lag time, growth rate) being evaluated.

Q	A	Decision Tree for Extrapolation (Kosher Chicken Skin)
1	no	Was the model validated for interpolation?
2	yes	Were data independent?
3	yes	Were data collected with the same methods as dependent data except for the new independent variable being evaluated?
4	yes	Were the independent variables at the same values as those used in model development?
5	yes	Was there a minimum of two prediction cases per combination of independent variables?
6	no	Did all combinations of the independent variables have the same number of prediction cases?
7	yes	Was the overall pAPZ ≥ 0.70?
8	no	Was pAPZ for all individual levels of independent variables ≥ 0.70?
9	yes	Was a single pAPZ ≥ 0.70 for every three consecutive combinations of the independent variables?
10	no	Was the model validated for extrapolation?

Average of APZ	Tim					
Temp	0	2	4	6	8	Total
5	1.00	0.85	1.00	0.80	1.00	0.93
10	1.00	0.95	0.85	0.64	1.00	0.89
15	1.00	0.95	1.00	0.95	0.95	0.97
20	1.00	1.00	1.00	0.79	0.10	0.78
25	1.00	1.00	0.93	0.84	0.81	0.92
30	1.00	1.00	0.87	0.70	0.46	0.81
35	1.00	0.75	1.00	1.00	1.00	0.95
40	1.00	0.84	0.98	0.86	1.00	0.93
45	1.00	0.66	0.74	0.48	0.50	0.68
50	1.00	1.00	0.50	0.50	0.75	0.75
Total	1.00	0.91	0.89	0.76	0.75	0.86

Fig. 5 Screenshot of the spreadsheet in the validation software tool (ValT) that has the decision tree and pivot table for independent data for extrapolation for primary, secondary, and tertiary models that predict log number. The decision tree has "yes" or "no" questions (Q) for test data criteria (Q2 to Q6), model performance criteria (Q7 to Q9), and model validation criteria (Q1 and Q10). The pivot table shows results for average acceptable prediction zones (APZ) values for individual combinations of independent variables (temperature, time), for individual levels of independent variables, and overall. This spreadsheet is similar to other versions of ValT for secondary models that predict lag time or growth rate

3.3 Criteria for Test Data

1. Dependent Data (Fig. 3)
   (a) Q1: Were the data used to develop the model?
      (i) If the data were not used for model development, this evaluation would not be valid since its purpose is to see how well the model predicts the data that were used to develop it.
   (b) Q2: Were the independent variables evenly spaced?
      (i) This criterion lessons the chance that the model will make biased predictions and it helps ensure that the evaluation of model performance is unbiased.
   (c) Q3: Was there a minimum of four prediction cases per combination of independent variables?
      (i) Replication is important for development of an accurate model and evaluation of its performance because it helps ensure that predictions are accurate and that an

important independent variable is not missing from the model. Four replications are the minimum. More is better.

(d) Q4: Did all combinations of independent variables have the same number of prediction cases?

(i) This criterion helps ensure that evaluation of model performance is unbiased.

2. Independent Data for Interpolation (Fig. 4)

(a) Q2: Were the data independent?

(i) This evaluation would not be valid if test data were used in model development. Thus, this criterion ensures that the model is evaluated against the proper data.

(b) Q3: Were the data collected using the same methods as used to collect dependent data?

(i) This criterion ensures that comparisons of observed and predicted values are not confounded by differences in data collection and modeling methods.

(ii) "Same methods" means same strain, inoculum size, previous history, food matrix, enumeration method, and modeling method.

(c) Q4: Were the independent variables at values intermediate to those used in model development?

(i) This criterion ensures even spacing of prediction cases and complete coverage of the interpolation region, which helps ensure an unbiased and complete evaluation of model performance for interpolation.

(d) Q5: Was there a minimum of two prediction cases per combination of independent variables?

(i) When combined with criteria for model performance (see below), this criterion ensures an accurate evaluation of model performance. Two replications are the minimum. More is better.

(e) Q6: Did all combinations of independent variables have the same number of prediction cases?

(i) This criterion helps ensure that evaluation of model performance is unbiased.

3. Independent Data for Extrapolation (Fig. 5)

(a) Q2: Were data independent?

(i) This criterion ensures that this evaluation is done against the proper type of data.

(b) Q3: Were data collected with the same methods as dependent data except for the new independent variable being evaluated?

   (i) This criterion when combined with a criterion for model validation (see below) helps ensure that comparisons of observed and predicted values are not confounded by differences in data collection and modeling methods. In addition, the model must be validated for dependent data and independent data for interpolation first, or these comparisons will be confounded.

   (ii) "Same methods" means same strain, inoculum size, previous history, food matrix, enumeration method, and modeling method unless one of these is the new independent variable being evaluated.

   (iii) Only one new independent variable can be evaluated at a time, or else a firm conclusion about model validation is not possible.

(c) Q4: Were the independent variables at the same values as those used in model development?

   (i) This criterion helps ensure an unbiased and complete evaluation of model performance.

   (ii) The same experimental design as used for model development is used so that if the model fails validation for extrapolation, these data can be used to develop a new model or expand the current model.

(d) Q5: Was there a minimum of two prediction cases per combination of independent variables?

   (i) This criterion helps ensure an accurate evaluation of model performance when coupled with criteria for model performance (see below).

   (ii) The idea is to use 50% less data for model validation than model development to save time and money.

   (iii) Two replications are the minimum, but more would be better.

(e) Q6: Did all combinations of independent variables have the same number of prediction cases?

   (i) This criterion helps ensure an unbiased evaluation of model performance.

3.4 Criteria for Model Performance

Criteria for model performance are the same for all three types of data.

1. Q5 or Q7: Was the overall pAPZ ≥0.70?
   (a) This criterion ensures that the evaluation for global prediction problems is objective.
   (b) This criterion says that the model must provide predictions with acceptable accuracy and bias at least 70% of the time. In arriving at this model performance metric, some predictions are considered fully acceptable, and some predictions are considered partly acceptable. The credit given for partly acceptable predictions decreases in a linear manner from the boundary of the corresponding fully acceptable prediction zones. The idea is to give partial credit for predictions as a function of how far they are from being fully acceptable.
   (c) The 70% acceptable value is based on the US education system where 70% correct answers on a test is considered acceptable or fair performance and on statistics where about 70% of observations in a normal distribution are within one standard deviation of the mean [18].
2. Q6 or Q8: Was pAPZ for all individual levels of independent variables ≥0.70?
   (a) This criterion ensures that the evaluation for local prediction problems is objective.
3. Q7 or Q9: Was a single pAPZ ≥0.70 for every three consecutive combinations of the independent variables?
   (a) This criterion ensures that this evaluation for local prediction problems is objective and is based on a minimum of 12 prediction cases for dependent data and 6 prediction cases for independent data for interpolation or extrapolation, which helps ensure that it is accurate.
   (b) This is an important step in the evaluation process because it isolates and evaluates specific phases in growth or death curves where responses change more rapidly, and prediction errors may be higher. Thus, it provides a stringent evaluation of model performance in the most sensitive regions of the prediction surface.

3.5 Criteria for Model Validation

1. Q8 (dependent) and Q1 (interpolation): Was the model validated for dependent data?
   (a) This criterion provides an objective evaluation of model validation for dependent data.
   (b) This criterion uses a formula (*see* **Note 5**) to automatically provide an answer of "yes" or "no."

2. Q10 (interpolation) and Q1 (extrapolation): Was the model validated for interpolation?
   (a) This criterion provides an objective evaluation of model validation for interpolation. A model cannot be validated for interpolation unless it was validated for dependent data.
   (b) This criterion uses a formula (*see* **Note 6**) to automatically provide an answer of "yes" or "no."
3. Q10 (extrapolation): Was the model validated for extrapolation?
   (a) This criterion ensures that this evaluation is objective.
   (b) A model cannot be validated for extrapolation unless it was validated for interpolation. Without validation for interpolation, comparison of observed and predicted values would be confounded, and this evaluation would not be valid.
   (c) This criterion uses a formula (*see* **Note 6**) to automatically provide an answer of "yes" or "no."

4 Example

To demonstrate the APZ method in ValT [24], a published tertiary model for growth (change in log number over time) of a low initial number (0.85 log) of *Salmonella enterica* serotype Typhimurium definitive phage type 104 (DT104) on chicken skin with native microflora as a function of time (0–8 h) and temperature (5–50 °C) and the corresponding secondary models for lag time and growth rate were evaluated [25]. The models were also evaluated for extrapolation to a new independent variable, which was a new food matrix (kosher chicken skin with native microflora). Kosher chicken skin differs from conventional chicken skin in that it is not subjected to scalding and it is exposed to kosher salt for 45 min during primary processing of the chicken; thus, it represents a different food matrix than conventionally processed chicken skin. The tertiary model was previously evaluated using an earlier version of the APZ method that did not have the same criteria for test data and did not evaluate the model for local prediction problems and did not have partly acceptable APZ, whereas the secondary models were not evaluated previously.

1. Tertiary model for log number
   (a) Dependent data (failed validation; Fig. 3)
      (i) Insufficient replication for one combination of time and temperature (Table 1)

Table 1
Pivot table results for prediction case count for the tertiary model for log number

Dependent	Temperature (°C)	Time (h) 0	2	4	6	8	Count
	5	*3*	*4*	*4*	*4*	*4*	*19*
	10	8	8	8	*7*	*7*	*38*
	15	8	8	8	8	8	40
	20	*7*	8	8	8	8	*39*
	25	*10*	*10*	*10*	*10*	*10*	*50*
	30	8	8	8	8	8	40
	35	8	8	8	8	8	40
	40	8	8	8	*7*	8	*39*
	45	8	8	8	8	*7*	*39*
	50	8	8	8	8	8	40
	Count	*76*	*78*	*78*	*76*	*76*	384
Interpolation	7.5	4	4	4	4	4	20
	12.5	4	4	4	4	4	20
	17.5	4	4	4	4	4	20
	22.5	4	4	4	4	4	20
	27.5	4	4	4	4	4	20
	32.5	4	4	4	4	4	20
	37.5	4	4	4	4	4	20
	42.5	4	4	4	4	*3*	*19*
	47.5	4	4	4	*3*	4	*19*
	Count	36	36	36	*35*	*35*	178
Extrapolation	5	4	4	4	4	4	20
	10	4	4	4	4	4	20
	15	4	4	4	4	4	20
	20	4	4	4	4	4	20
	25	4	4	4	4	4	20
	30	4	4	4	4	4	20
	35	4	4	4	4	4	20
	40	4	4	*3*	4	*3*	*18*
	45	4	*3*	*3*	4	4	*18*
	50	4	4	4	4	4	20
	Count	40	*39*	*38*	40	*39*	196

Table 2
Pivot table results for average APZ value (pAPZ) for the tertiary model for log number

		Time (h)					
Dependent	**Temperature (°C)**	**0**	**2**	**4**	**6**	**8**	**Average**
	5	1.00	0.95	0.85	0.85	1.00	0.92
	10	1.00	0.98	0.82	0.97	0.88	0.93
	15	1.00	0.82	0.97	0.95	0.95	0.94
	20	1.00	0.97	0.85	0.90	*0.49*	0.84
	25	1.00	0.97	1.00	0.80	0.97	0.95
	30	1.00	1.00	0.97	0.89	0.93	0.96
	35	1.00	1.00	1.00	1.00	0.94	0.99
	40	1.00	1.00	0.99	0.84	0.97	0.96
	45	1.00	0.95	0.99	0.84	0.73	0.91
	50	1.00	0.85	*0.57*	*0.67*	*0.65*	0.75
	Average	1.00	0.95	0.91	0.87	0.85	0.91
Interpolation	7.5	1.00	1.00	0.95	1.00	0.95	0.98
	12.5	1.00	0.84	0.90	0.95	1.00	0.94
	17.5	1.00	0.95	1.00	0.90	0.80	0.93
	22.5	1.00	1.00	0.98	1.00	0.89	0.97
	27.5	1.00	0.98	1.00	1.00	*0.46*	0.89
	32.5	1.00	0.98	1.00	0.98	0.97	0.99
	37.5	1.00	1.00	1.00	0.93	0.97	0.98
	42.5	1.00	0.88	1.00	0.81	0.72	0.89
	47.5	1.00	1.00	1.00	0.82	*0.56*	0.88
	Average	1.00	0.96	0.98	0.93	0.81	0.94
Extrapolation	5	1.00	0.85	1.00	0.80	1.00	0.93
	10	1.00	0.95	0.85	*0.64*	1.00	0.89
	15	1.00	0.95	1.00	0.95	0.95	0.97
	20	1.00	1.00	1.00	0.79	*0.10*	0.78
	25	1.00	1.00	0.93	0.84	0.81	0.92
	30	1.00	1.00	0.87	0.70	*0.46*	0.81
	35	1.00	0.75	1.00	1.00	1.00	0.95
	40	1.00	0.84	0.98	0.86	1.00	0.93
	45	1.00	0.66	0.74	*0.48*	*0.50*	*0.68*
	50	1.00	1.00	*0.50*	*0.50*	0.75	0.75
	Average	1.00	0.91	0.89	0.76	0.75	0.86

(ii) Uneven distribution of data among combinations of time and temperature (Table 1)

(iii) Three consecutive pAPZ <0.7 (Table 2)

(b) Interpolation (failed validation; Fig. 4)

(i) Failed validation for dependent data (Fig. 3)

(ii) Values for time were not at intermediate values (Table 1)

(iii) Uneven distribution of data (Table 1)

(c) Extrapolation (failed validation; Fig. 5)

(i) Failed validation for interpolation (Fig. 4)

(ii) Uneven distribution of data (Table 1)

(iii) pAPZ for one temperature was <0.7 (Table 2)

2. Secondary model for lag time (Table 3)

(a) Dependent data (failed validation)

(i) Insufficient replication

(ii) Uneven distribution of data

(iii) pAPZ <0.7 for one temperature

(b) Interpolation (failed validation)

(i) Failed validation for dependent data

(ii) pAPZ <0.7 for three temperatures

(c) Extrapolation (failed validation)

(i) Failed validation for interpolation

(ii) Overall pAPZ <0.70

(iii) pAPZ <0.70 for four temperatures

3. Secondary model for growth rate (Table 3)

(a) Dependent data (failed validation)

(i) Insufficient replication

(ii) Uneven distribution of data

(iii) pAPZ <0.7 for three temperatures

(b) Interpolation data (failed validation)

(i) Failed validation for dependent data

(ii) Overall pAPZ <0.7

(iii) pAPZ <0.7 for four temperatures

(iv) pAPZ <0.7 for three consecutive temperatures

(c) Extrapolation data (failed validation)

(i) Failed validation for interpolation

(ii) overall pAPZ <0.7

(iii) pAPZ <0.7 for five temperatures

(iv) pAPZ <0.7 for three consecutive temperatures

Table 3
Pivot table results for prediction case count and average APZ value (pAPZ) for the secondary models for lag time and growth rate

		Lag time		Growth rate	
Dependent	**Temperature (°C)**	**Count**	**Average**	**Count**	**Average**
	5	2	1.00	2	1.00
	10	4	1.00	4	1.00
	15	4	1.00	4	1.00
	20	4	1.00	4	*0.00*
	25	5	0.85	5	*0.41*
	30	4	1.00	4	1.00
	35	4	1.00	4	1.00
	40	4	1.00	4	1.00
	45	4	0.75	4	*0.57*
	50	4	*0.00*	4	1.00
	Overall	39	0.85	39	0.78
Interpolation	7.5	2	1.00	2	1.00
	12.5	2	1.00	2	1.00
	17.5	2	1.00	2	*0.00*
	22.5	2	*0.50*	2	*0.50*
	27.5	2	*0.50*	2	*0.00*
	32.5	2	1.00	2	1.00
	37.5	2	1.00	2	1.00
	42.5	2	1.00	2	1.00
	47.5	2	*0.00*	2	*0.44*
	Overall	18	0.78	18	*0.66*
Extrapolation	5	2	1.00	2	1.00
	10	2	1.00	2	1.00
	15	2	1.00	2	1.00
	20	2	1.00	2	*0.00*
	25	2	*0.62*	2	*0.50*
	30	2	0.91	2	*0.69*
	35	2	1.00	2	0.75
	40	2	0.89	2	1.00
	45	2	*0.02*	2	*0.33*
	50	2	*0.00*	2	*0.00*
	Overall	20	0.74	20	*0.63*

5 Conclusion

1. Model repair
   (a) The purpose of model validation is not to reject models. Rather it is to identify prediction problems and fix them before the model is provided to end users. The model used in the above example was repaired, published, and distributed to end users [24]. Specifically,
      (i) Data were culled to address insufficient replication, uneven distribution of data, and local prediction problems. The result was a new model with reduced temperature range (10 to 40 °C).
      (ii) Missing data were estimated within a storage trial using a primary model to address uneven distribution of data.
      (iii) A new model was used. The tertiary model was replaced with a neural network that was developed in one step and eliminated the secondary models for lag time and growth rate and resulted in a model with better global and local performance.
      (iv) The new model was validated for dependent data, interpolation, and extrapolation to kosher chicken skin.

6 Notes

1. Residual (log number) is calculated in ValT as follows:
   (a) IF(observed = ““, ““,IF(observed*predicted<0,-2.01, observed-predicted)) where
      (i) If the cell for the observed log number is blank, the cell for the corresponding residual is blank; otherwise,
         1. ValT is programmed for 500 prediction cases. If the dataset has less than 500 prediction cases, this command ensures rows with no data are ignored in calculations.
            (a) ValT can be expanded to handle more than 500 prediction cases. Formula for calculating residuals and assigning APZ values can be copy and pasted into additional cells of the spreadsheet. In addition, the data source for the pivot table can be manually reset to cover the new range of data.

(b) ValT can also be expanded to have more independent variables. Columns for additional independent variables can be inserted to the left of the column for observed values without the need to manually adjust cell addresses in formula used to calculate residuals and assign APZ values. However, the data source for the pivot table would need to be manually reset to account for the new range of data.

(ii) If the product of the observed and corresponding predicted log number is less than zero, the default residual is −2.01 log; otherwise,

1. This command ensures proper assignment of APZ values for special prediction cases, which are prediction cases for which the model predicts survival, but no survival is observed or for which the model predicts no survival, but survival is observed. When the pathogen is not detected or predicted to not be present (no survival), a default value of −0.01 log is assigned based on the convention used in ComBase, an international microbial modeling database [26]. Thus, if there is disagreement between observed and predicted log numbers, the product will be a negative number, and a default residual of −2.01 is assigned that is outside the APZ and will result in assignment of the appropriate APZ value of zero. On the other hand, if there is agreement between observed and predicted values, the product of the values will be greater than or equal to zero and the residual will be calculated.

2. In the special case where both observed and predicted values are −0.01 because the pathogen was not detected or predicted not to be there, the residual will be calculated and will equal 0, and the appropriate APZ value of 1 will be assigned.

(iii) The residual is equal to the observed value minus the predicted value.

1. Negative residuals are fail-safe and positive residuals are fail-dangerous.

2. Relative residual (proportion) is calculated in ValT using the following formula:

(a) =IF(observed = ““, ““,IF(observed-predicted = 0,0,IF (predicted = 0,-1.21, ((predicted-observed)/ predicted)))) for lag time models

(b) =IF(observed = "", "",IF(observed-predicted = 0,0,IF (predicted = 0,-0.61,((observed-predicted)/predicted)))) for growth rate models where

   (i) If the cell for the observed value is blank, the cell for the corresponding relative residual and APZ value are blank; otherwise,

   (ii) if the difference between the observed and predicted values is zero, the relative residual is zero (perfect agreement); otherwise,

   (iii) if the predicted value for lag time or growth rate is zero and the observed value for lag time or growth rate is nonzero, then the relative residual is a default negative value for lag time (−1.21) and a default negative value for growth rate (−0.61); otherwise,

      1. These commands ensure proper assignment of APZ values for special prediction cases.

      2. Any negative number outside the APZ can be used.

   (iv) The relative residual is equal to the predicted value minus the corresponding observed value divided by the corresponding predicted value for lag time models or the observed value minus the corresponding predicted value divided by the corresponding predicted value for growth rate models.

      1. For both lag time and growth rate models, negative relative residuals are fail-safe and positive relative residuals are fail-dangerous.

      2. Although it is common practice to use the observed value in the denominator when making these calculations [16], the predicted value is used here because the relative distance of the residual from the model prediction is of greater interest in predictive microbiology for foodborne pathogens.

3. Acceptable prediction zones (APZ) are twice as wide in the fail-safe direction because when a model is used to predict food safety, it can err more in the fail-safe direction to provide an added level of safety [1]. However, it is important that a model not err too much in either the fail-safe or fail-dangerous direction; otherwise, safe food would be removed from the food supply, and unsafe food would not be removed from the food supply, respectively. Nonetheless, there are four APZ: two in the fail-safe direction and two in the fail-dangerous direction. Boundaries of APZ differ by model type.

(a) Primary, secondary, or tertiary model for log number
   (i) > −2 log to < −1 log (fail-safe and partly acceptable)
   (ii) −1 log to ≤0 log (fail-safe and fully acceptable)
   (iii) ≥ 0 log to 0.5 log (fail-dangerous and fully acceptable)
   (iv) > 0.5 log to <1 log (fail-dangerous and partly acceptable)
      1. Residuals (observed – predicted) in APZ i and iv (partly acceptable predictions) are assigned a value from >0 to <1 for calculation of pAPZ, whereas residuals in APZ ii and iii (fully acceptable predictions) are assigned a value of one for calculation of pAPZ. Residuals outside the APZ (unacceptable predictions) are assigned a value of zero for calculation of pAPZ, which is the proportion of residuals in the four APZ and is equal to the sum of assigned APZ values divided by the total number of APZ values or is equal to the average of APZ values.

(b) Secondary model for lag time (h)
   (i) > −1.2 to < −0.6 (fail-safe and partly acceptable)
   (ii) −0.6 to ≤0 (fail-safe and fully acceptable)
   (iii) ≥ 0 to 0.3 (fail-dangerous and fully acceptable)
   (iv) > 0.3 to <0.6 (fail-dangerous and partly acceptable)
      1. Relative residuals ((predicted – observed)/predicted)) in APZ i and iv (partly acceptable predictions) are assigned an APZ value from >0 to <1 for calculation of pAPZ, whereas relative residuals in APZ ii and iii (fully acceptable predictions) are assigned an APZ value of one for calculation of pAPZ. Relative residuals outside the APZ (unacceptable predictions) are assigned an APZ value of zero for calculation of pAPZ.
      2. The APZ for lag time are twice as wide as APZ for growth rate because lag time is more variable than growth rate due to greater variation of lag time among individual cells [18].

(c) Secondary model for growth rate (log/h)
   (i) > −0.6 to < −0.3 (fail-safe and partly acceptable)
   (ii) −0.3 to ≤0 (fail-safe and fully acceptable)
   (iii) ≥ 0 to 0.15 (fail-dangerous and fully acceptable)
   (iv) > 0.15 to <0.3 (fail-dangerous and partly acceptable)

1. Relative residuals ((observed – predicted)/predicted)) in APZ i and iv (partly acceptable predictions) are assigned an APZ value from >0 to <1 for calculation of pAPZ, whereas relative residuals in APZ ii and iii (fully acceptable predictions) are assigned an APZ value of 1 for calculation of pAPZ. Relative residuals outside the APZ (unacceptable predictions) are assigned an APZ value of 0 for calculation of pAPZ.

4. APZ values in the APZ method are assigned in ValT as follows:
   (a) For a primary, secondary, or tertiary model that predicts log number
      (i) APZ = 0 if the residual is ≤ −2 log (fail-safe) or ≥ 1 log (fail-dangerous).
      (ii) APZ = 1 if the residual is ≥ −1 log (fail-safe) to ≤0.5 log (fail-dangerous).
      (iii) APZ = 1 log – ((residual +1 log)/−1 log) if the residual is > – 2 log (fail-safe) and < −1 log (fail-safe).
      (iv) APZ = 1 – ((residual – 0.5 log)/0.5 log) if the residual is >0.5 log (fail-dangerous) and < 1 log (fail-dangerous).
   (b) For a secondary model that predicts lag time (h)
      (i) APZ = 0 if the relative residual is ≤ −1.2 (fail-safe) or ≥ 0.6 (fail-dangerous).
      (ii) APZ = 1 if the relative residual is ≥ −0.6 (fail-safe) to ≤0.3 (fail-dangerous).
      (iii) APZ = 1 – ((relative residual +0.6)/−0.6) if the relative residual is > – 1.2 (fail-safe) and < −0.6 (fail-safe).
      (iv) APZ = 1 – ((relative residual – 0.3)/0.3) if the relative residual is >0.3 (fail-dangerous) and < 0.6 log (fail-dangerous).
   (c) For a secondary model that predicts growth rate (log/h)
      (i) APZ = 0 if the relative residual is ≤ −0.6 (fail-safe) or ≥ 0.3 (fail-dangerous).
      (ii) APZ = 1 if the relative residual is ≥ −0.3 (fail-safe) to ≤0.15 (fail-dangerous).
         1. This APZ was based on an established APZ for generation time [1].
      (iii) APZ = 1 – ((relative residual +0.3)/−0.3) if the relative residual is > – 0.6 (fail-safe) and < −0.3 (fail-safe).

(iv) APZ = 1 – ((relative residual – 0.15)/0.15) if the relative residual is >0.15 (fail-dangerous) and < 0.3 log (fail-dangerous).

5. Instructions for evaluating dependent data for compliance with test data, model performance, and model validation criteria of the APZ method in ValT are found in spreadsheet 3 and begin on row 36 and are as follows:

   (a) To evaluate the model for test data criteria, update results in the pivot table as follows:

      (i) Click on cell A10 and then click on Pivot Table in the toolbar, and then click on Analyze and then click on Refresh to update the data analysis for counting prediction cases per combination and level of independent variables and overall. However, it is possible that results might not appear properly for two reasons: first, because previous data had different independent variables and(or) different values for independent variables, and, second, because the previous evaluation was for pAPZ. To fix these issues, the following steps would be taken:

         1. To fix the first problem, click on the Filters for Rows and Columns one at a time, and then select only those values for independent variables that are relevant for the current analysis.

         2. To fix the second problem, click on Field List in the toolbar and then in the Field List click on Σ Values Pull Down Menu and then click on Value Field Settings and then click on Select Count.

      (ii) Answer (A) questions (Q) 1 to 4 for test data criteria in the Decision Tree by selecting "yes" or "no" from the pull-down menus.

   (b) To evaluate the model for model performance criteria, do the following:

      (i) Change Σ Value from Count to Average by clicking on Field List, Σ Values Pull-Down Menu, and then click on Value Field Settings and then click on Select Average.

      (ii) Click on any pAPZ value in the pivot table and then click on APZ Macro button to highlight (red fill) pAPZ that are <0.7. However, this may happen automatically in which case this step does not have to be performed.

      (iii) Answer (A) questions (Q) 5 to 7 for model performance criteria in the Decision Tree by selecting "yes" or "no" from the pull-down menu.

(c) To evaluate the model for validation criteria, nothing needs to be done because the spreadsheet contains a formula for question 8 that provides an automatic answer of "yes" or "no" based on the answers to questions 1 to 7:

(i) Q8 B9 = IF(CONCATENATE(B2,B3,B4,B5,B6,B7, B8) = 'yesyesyesyesyesyesyes', 'yes','no')

1. Where if answers to questions 1 to 7 in cells B2 to B8 are "yes," then the answer is "yes"; otherwise, it is "no."

(a) This answer is linked to cell B2 or Q1 in spreadsheet 5 because validation for dependent data is a requirement for validation for interpolation.

6. Instructions for evaluating independent data for interpolation or independent data for extrapolation for compliance with test data, model performance, and model validation criteria of the APZ method in ValT are found in spreadsheet 5 and begin on row 36 for independent data for interpolation and in spreadsheet 7 and begin on row 36 for independent data for extrapolation. The instructions are like those for dependent data, with the following exceptions:

(a) To evaluate the model for test data criteria, update results in the pivot table as described above for dependent data, except answer (A) questions (Q) 2 to 6 for test data criteria in the Decision Tree, by selecting "yes" or "no" from the pull-down menus.

(b) To evaluate the model for performance criteria, follow directions for dependent data, except answer (A) questions (Q) 7 to 9 for model performance criteria in the Decision Tree, by selecting "yes" or "no" from the pull-down menu.

(c) To evaluate the model for validation criteria, nothing needs to be done because the spreadsheets contain formula for Q1 and Q10 that provide an automatic answer of "yes" or "no" based on the answer to the last question in the previous step and the answers to Q1 to Q9 in the current step, respectively. The formula are as follows:

1. Q1:'5'!B2 = '3'B9 or '7'!B2 = '5'!B11).
2. Q10:'5'!B11 or '7'!B11 = IF(CONCATENATE(B2, B3,B4,B5,B6,B7,B8,B9,B10)='yesyesyesyesyesyesyesyesyes','yes','no') where if all the answers to questions 1 to 9 in cells B2 to B10 are "yes," then the answer is "yes"; otherwise, the answer is "no."

7. The formula used for assigning APZ values are as follows:

(a) For primary, secondary, and tertiary models that predict log number,

(i) IF(observed-predicted = ““,““,IF(observed-predicted<= − 2,0,IF(observed-predicted> = 1,0, VLOOKUP(observed-predicted,’8′!$A$2:$B $1503,2,TRUE)))) where

1. If the corresponding cell for the residual (observed − predicted) is blank, the cell for the APZ value is blank; otherwise,
2. If the residual (observed − predicted) is less than or equal to −2 log, the APZ value is zero; otherwise,
3. If the residual (observed − predicted) is greater than or equal to 1 log, the APZ value is 0; otherwise,
4. The residual (observed − predicted) is used to vertically lookup (VLOOKUP) an APZ value from >0 to 1 in column 2 of an array (A2:B1503) in spreadsheet 8 (‘8’!) where the match must be “True,” which means the closet match.

(b) For secondary models that predict lag time,

(i) =IF((predicted − observed)/predicted = ““,““,IF ((predicted − observed)/predicted <= − 1.2,0,IF ((predicted − observed)/predicted > = 0.6,0, VLOOKUP((predicted − observed)/predicted,’8′! $A$2:$B$903,2,TRUE)))) where

1. If the corresponding cell for the relative residual ((predicted − observed)/predicted) is blank, the cell for the APZ value is blank; otherwise,
2. If the relative residual ((predicted − observed)/predicted) is less than or equal to −1.2, the APZ value is 0; otherwise,
3. If the relative residual ((predicted − observed)/predicted) is greater than or equal to 0.6, the APZ value is 0; otherwise,
4. The relative residual ((predicted − observed)/predicted) is used to vertically lookup (VLOOKUP) an APZ value from >0 to 1 in column 2 of an array (A2:B903) in spreadsheet 8 (‘8’!) where the match must be “True,” which means the closet match.

(c) For secondary models that predict growth rate,

(i) =IF((observed – predicted)/predicted = "","",IF ((observed – predicted)/predicted <= – 0.6,0,IF ((observed – predicted)/predicted > = 0.3,0, VLOOKUP((observed – predicted)/predicted,'8'! $A$2:$B$453,2,TRUE))))

1. Where, if the corresponding cell ((observed – predicted)/predicted) for the relative residual is blank, the cell for the APZ value is left blank; otherwise,
2. If the relative residual ((observed – predicted)/predicted) is less than or equal to –0.6, the APZ value is 0; otherwise,
3. If the relative residual ((observed – predicted)/predicted) is greater than or equal to 0.3, the APZ value is 0; otherwise,
4. The relative residual ((observed – predicted)/predicted) is used to vertically lookup (VLOOKUP) the APZ value from >0 to 1 in column 2 of an array (A2:B453) in spreadsheet 8 ('8'!) where the match must be "True," which means the closet match.

References

1. Ross T (1996) Indices for performance evaluation of predictive models in food microbiology. J Appl Bacteriol 81:501–508
2. Baranyi J, Pin C, Ross T (1999) Validating and comparing predictive models. Int J Food Microbiol 48:159–166
3. Ross T, Dalgaard P, Tienungoon S (2000) Predictive modeling of the growth and survival of *Listeria* in fishery products. Int J Food Microbiol 62:231–245
4. Walls I, Scott VN (1996) Validation of predictive mathematical models describing the growth of *Escherichia coli* O157:H7 in raw ground beef. J Food Prot 59:1331–1335
5. Mejlholm O, Gunvig A, Borggaard C, Blom-Hanssen J, Mellefont L, Ross T, Leroi F, Else T, Visser D, Dalgaard P (2010) Predicting growth rates and growth boundary of *Listeria monocytogenes* – an international validation study with focus on processed and ready-to-eat meat and seafood. Int J Food Microbiol 141(3):137–150
6. Whiting RC, Buchanan RL (1993) A classification of models for predictive microbiology – a reply to K. R. Davey. Food Microbiol 10:175–177
7. Baranyi J, Roberts TA, McClure P (1993) A non-autonomous differential equation to model bacterial growth. Food Microbiol 10: 43–59
8. Baranyi J, Roberts TA (1994) A dynamic approach to predicting bacterial growth in food. Int J Food Microbiol 23:277–294
9. Ratkowsky DA, Olley J, McMeekin TA, Ball A (1982) Relationship between temperature and growth rate of bacterial cultures. J Bacteriol 149:1–5
10. Ratkowsky DA, Lowry RK, McMeekin TA, Stokes AN, Chandler RE (1983) Model for bacterial culture growth rate throughout the entire biokinetic temperature range. J Bacteriol 154:1222–1226
11. Buchanan RL, Phillips JG (1990) Response surface model for predicting the effects of temperature, pH, sodium chloride content, sodium nitrite concentration and atmosphere on the growth of *Listeria monocytogenes*. J Food Prot 53:370–376
12. Buchanan RL, Bagi LK, Goins RV, Phillips JG (1993) Response surface models for the growth kinetics of *Escherichia coli* O157:H7. Food Microbiol 10:303–315
13. Buchanan RL (1991) Using spreadsheet software for predictive microbiology applications. J Food Saf 11:123–134
14. McClure PJ, Blackburn CW, Cole MB, Curtis PS, Jones JE, Legan JD, Ogden ID, Peck MW,

Roberts TA, Sutherland JP, Walker SJ (1994) Modelling the growth, survival and death of microorganisms in foods: the UK Food Micromodel approach. Int J Food Microbiol 23: 265–275

15. Sutherland JP, Aherne A, Beaumont AC (1996) Preparation and validation of a growth model for *Bacillus cereus*: the effects of temperature, pH, sodium chloride and carbon dioxide. Int J Food Microbiol 30:359–372
16. Delignette-Muller ML, Rosso L, Flandrois JP (1995) Accuracy of microbial growth predictions with square root and polynomial models. Int J Food Microbiol 27:139–146
17. Zwietering MH, Cuppers HGAM, de Wit JC, van Riet TK (1994) Evaluation of data transformations and validation of a model for the effect of temperature on bacterial growth. Appl Environ Microbiol 60:195–203
18. Oscar TP (2005) Validation of lag time and growth rate models for *Salmonella* Typhimurium: acceptable prediction zone method. J Food Sci 70:M129–M137
19. Oscar TP (2005) Development and validation of primary, secondary and tertiary models for predicting growth of *Salmonella* Typhimurium on sterile chicken. J Food Prot 68:2606–2613
20. Min KJ, Yoon KS (2010) Development and validation of a predictive model for foodborne pathogens in ready-to-eat pork as a function of temperature and a mixture of potassium lactate and sodium diacetate. J Food Prot 73(9): 1626–1632
21. Mishra A, Guo M, Buchanan RL, Schaffner DW, Pradhan AK (2016) Development of growth and survival models for *Salmonella* and *Listeria monocytogenes* during non-isothermal time-temperature profiles in leafy greens. Food Control 71:32–41
22. Juneja VK, Mishra A, Pradhan AK (2018) Dynamic predictive model for growth of *Bacillus cereus* from spores in cooked beans. J Food Prot 81(2):308–315. https://doi.org/10.4315/0362-028X.JFP-17-391
23. Mohr TB, Juneja VK, Thippareddi HH, Schaffner DW, Bronstein PA, Silverman M, Cook LVJ (2015) Assessing the performance of *Clostridium perfringens* cooling models for cooked, uncured meat and poultry products. J Food Prot 78:1512–1526
24. Oscar TP (2020) Validation software tool (ValT) for predictive microbiology based on the acceptable prediction zones method. Int J Food Sci Technol 55:2802–2812
25. Oscar TP (2009) Predictive model for survival and growth of *Salmonella* Typhimurium DT104 on chicken skin during temperature abuse. J Food Prot 72:304–314
26. Baranyi J, Tamplin ML (2004) ComBase: a common database on microbial responses to food environments. J Food Prot 67:1967–1971

Chapter 11

Predictive Modeling for Spoilage Microorganisms

Çağla Pınarlı and Fatih Tarlak

Abstract

Every year, tons of food is thrown away because of changes in odor, taste, texture, or color. Food spoilage has a very important effect at this point. Microorganisms have a significant position in food spoilage. The type of microorganisms seen varies depending on factors such as the amount of water, temperature, acidity, carbohydrate, protein, or fat ratios of foods, packaging type, and oxygen levels. Temperature is the most important one for spoilage. In this section, a basic organism that causes food spoilage in different food groups, *Pseudomonas* spp., has been specifically reviewed. General modeling of microbial degradation and basic principles of microbial experiments are also given in that part.

Key words *Pseudomonas*, Shelf life, Secondary model, Baranyi model

1 Introduction

Food is required for humans to grow, develop, maintain, and improve their health, as well as to fulfill their daily macro- and micronutrient needs. There may be some difficulties in obtaining nutritious and appetizing foods at this time. One of these issues is food spoilage [1]. Food spoilage is a metabolic process that results in changes in sensory qualities, which have a detrimental impact on food intake. Foods that have deteriorated may be consumed safely; in other words, there are no germs or poisons present; therefore, eating them won't get you sick. On the other hand, food spoilage results in the rejection of nutrients due to changes in texture, smell, taste, or appearance. This scenario is assumed to be caused mostly by microorganisms. As a result, the animals reject the nutrients, and the bacteria keep the food supply to themselves [2].

During storage and distribution, food decomposition caused by spoilage microorganisms has a significant influence on food quality and shelf life. Food spoiling is defined as a process of change in a product that renders it unfit for ingestion [3]. Food spoiling is induced by physical and bio(chemical) interactions, as well as parasite activity. The underlying mechanisms are extremely complicated

Verônica Ortiz Alvarenga (ed.), *Basic Protocols in Predictive Food Microbiology*, Methods and Protocols in Food Science,
https://doi.org/10.1007/978-1-0716-3413-4_11,

and difficult to uncover; therefore, they are still a mystery [4]. However, the microbiological proliferation of spoilage organisms and the impacts of their extracellular enzymes, such as produced proteases, lipases, and carbohydrases, are primarily responsible for the loss of desired qualities and considerable product variations. At all phases of production (harvesting, processing, and storage), food can be infected with a vast range of microbes, although only a tiny percentage of them can grow into food products, causing major spoilage. Visible yeast or mold colonies, gas production, turbidity of liquids, discoloration, and changes in distinctive taste and aroma are all signs of microbial deterioration [5].

Microorganisms that contaminate food and cause foodborne disease are studied in food microbiology. Microbes are present in the foods we eat, and they are seldom sterile. Microbial loads are retained in foods, and their composition is very varied. Microorganisms are found in raw materials' common microflora, as well as during animal slaughter, food harvesting, processing, storage, and distribution [6, 7]. In the vast majority of cases, the food is ingested without issue and with no negative health consequences [7].

More than one-third of all food meant for human consumption is wasted or spoiled before it comes to the consumer, according to the FAO (Food and Agriculture Organization of the United Nations) [8]. Food losses from farm to fork have environmental and economic impacts [9]. According to the Food Waste Index 2021 report, 17% of the food produced is wasted before it reaches the consumer. In this case, it is thought that at least 40% of the food produced is wasted. In numerical terms, it is thought that 2.5 billion tons of food are lost or wasted annually. Forty-four percent of these foods are fish and seafood; 26% fruits and vegetables; 15% roots, tubers, and oil crops; 14% cereals and pulses; 12% meat and animal products; and 6% other foods [10]. At this point, the importance of food losses in fighting hunger and improving food security comes to the fore, especially in poor countries around the world. It is affected by factors such as food production conditions, transportation, distribution conditions, and infrastructure. These preventable food losses directly or indirectly influence the financial situation of both farmers and consumers negatively. Food spoilage and loss also have environmental effects. When a certain percentage of the product is lost, the fertilizer, water, and energy used to grow that crop are also wasted at the same rate [5]. While the water footprint spent for the production of 1 kilogram of red meat is more than 15,000 liters on average, the water used for the production of 1 kg of eggs is 3265 liters, and the water footprint for 1 kg of wheat is 1827 liters [11].

Food predictive microbiology is a vast scientific subject that encompasses a variety of principles and applications. Readers who are unfamiliar with this vocabulary may find it difficult to follow the language employed by researchers in this field. The predictive microbiology framework is introduced to new predictive microbiology practitioners and researchers in this part. Predictive microbiology is a scientific area of the subject of food microbiology that aims to statistically characterize microbial activity in food settings in order to develop appropriate mathematical models. A mathematical simple implementation of a real system based on its more significant aspects that is described through mathematical equations [12].

Predictive models are useful tools for determining product shelf life and food safety, doing hazard analysis and establishing crucial control points, and developing risk assessment programs. Predictive models are a rapid, efficient, and cost-effective approach of analyzing the potential for growth or forecasting the development of microorganisms under certain conditions without the need for lengthy, expensive, and time-consuming analysis procedures or specialist personnel. Traditional microbial counting approaches may be used to evaluate microbial load in meals; however, these techniques are time consuming and expensive, and they require expert personnel. Furthermore, these approaches provide information only for the conditions under which the analysis was performed; therefore, using classic enumerated techniques to determine the growth of microorganisms under fluctuating environmental factors is not viable. Predictive models, on the other hand, can forecast the development of microbes in foods throughout processing and storage in actual time [13].

The duration to attain a defined goal level under varying temperature circumstances was the most critical parameter in predictive models used to characterize the growth behavior of spoilage bacteria. Modeling the growth behavior of *Pseudomonas* spp., one of the most common bacteria isolated from a variety of foods, is critical for determining the best storage conditions for food products and predicting their shelf life. The duration to attain a defined goal level under varied temperature circumstances is the most significant parameter in predictive models used to characterize the growth behavior of spoilage bacteria. *Pseudomonas* spp., one of the most common microbes extensively identified from a variety of foods, has been investigated and simulated using certain prediction models [14]. Koutsoumanis utilized the logistic model to characterize the proliferation of *Pseudomonas* spp. in fish maintained at temperatures varying from 0 to 15 °C [15]. Gospavic et al. implemented a growth model for predicting the development of *Pseudomonas* spp. in chicken under varied temperature circumstances using modified Gompertz and Baranyi models [16]. Zhang et al. utilized the Baranyi model to fit the amount of *Pseudomonas* spp. in beef held between 0 and 20 °C [17]. Bruckner et al. utilized the modified Gompertz model to determine *Pseudomonas* spp. growth behavior in pig and poultry meat refrigerated between 2 and 15 °C [18]. To explain the growth behavior of *Pseudomonas* spp. in tropical fresh shrimp held at temperatures ranging from 0 to 28 °C, Dabadé et al. employed the Baranyi and modified Gompertz models [19]. Lytou et al. employed the Baranyi model to characterize the growth of total viable bacteria, including *Pseudomonas* spp., in marinated and unmarinated chicken breast fillets held at 4, 10, and 15 °C and connected this data with shelf life [20]. Psychrotrophic bacteria, mostly *Pseudomonas* spp., have been identified as the most common microbes responsible for mushroom deterioration [21]. The most critical element impacting rate of growth is temperature.

2 Materials and Methods

Modeling activities in the field of predictive microbiology are often classified into four key phases, as illustrated in Fig. 1. The next subsections go into the specifics of the process.

Food products are procured and delivered to the laboratory where microbiological studies are conducted for general microbial deterioration experiments. Environmental variables that directly impact the development behavior of spoilage microorganisms in food should be kept under control during transit. After the food goods arrive at the lab, initial microbial counts may be obtained, and the data generating procedure for microbial deterioration can begin. To track and simulate microbial deterioration, it's critical to

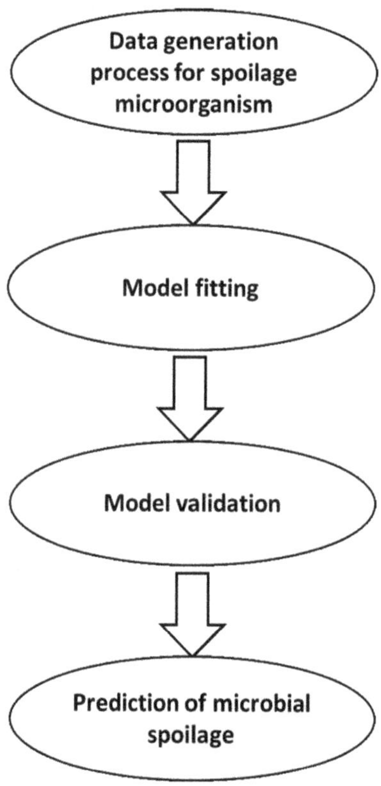

Fig. 1 General steps for modeling of microbial spoilage of food products

determine the starting microbial population accurately. To track and simulate microbial deterioration, it is critical to accurately determine the starting microbial population. As a result, food samples should be picked in a homogeneity-effective manner, with samples reflecting the microbiological quality of the entire batch of food items.

3 Microbial Experiments

Pseudomonas spp. is one of the most common bacteria species in food and is known to cause spoiling in a variety of foods [22]. *Pseudomonas* spp. determination growth behavior is widely researched with the goal of simulating microbial deterioration in food goods. As a result, the experimental technique will be focused on describing the microbiological growth behavior of *Pseudomonas* spp. Some food product (generally 25 g or mL food product) portions are aseptically weighed in a sterile Stomacher bag containing 225 mL of sterile Ringer's solution and homogenized in a Stomacher apparatus for sufficient time to obtain a homogeneous food/solution Ringer's mixture for microbiological experiments. For the enumeration of naturally existing *Pseudomonas* spp. in food items, appropriate serial decimal dilutions in Ringer's solution were surface plated on *Pseudomonas* agar base supplemented with CFC (cephalothin, fucidin, cetrimide) selective supplement [23]. To accurately quantify microbe counts, experimental duplicates are required, and at least three separate experiments are recommended for microbiological examination (see more information in Chap. 3).

4 Fitting the Model

The mathematical models that are employed in the model fitting stage are grouped into three categories: primary, secondary, and tertiary models [24]. Under a steady environmental situation, primary models describe growth data as a function of time. Models of the sigmoidal type, such as the modified Gompertz, logistic, Baranyi, and Huang models, are commonly used to fit microbial growth data. Secondary models show how environmental elements like temperature, pH, and water activity (a_w) affect basic model parameters like maximal specific growth rate and lag phase duration. Temperature is one of the most critical environmental elements in terms of food safety. The Ratkowsky or square root model (see more information in Chap. 1) is the most often used secondary model for determining the link between temperature and maximal specific growth rate [25]. To analyze microbiological behavior under specific conditions, tertiary models integrate both primary and secondary models with user-friendly application software or expert systems [24, 26].

4_C - Notepad

File Edit Format View Help

Time	conc	Temp
0	3.7	4
44	4.9	4
89	3.8	4
114	4	4
139	4	4
187	4.3	4
259	4.7	4
330	5.3	4
380	6.1	4
428	6.3	4
501	7.4	4
548	7.6	4
599	8	4
646	7.8	4
694	7.9	4
738	7.5	4
790	7.5	4

Ln 1, Col 1 100% Windows (CRLF) UTF-8

Fig. 2 Example of dataset for fit Baranyi primary growth model and square root secondary model

For analyzing *Pseudomonas* behavior under specific temperature conditions, they may use Baranyi model (see equation in Chap. 4) for estimating the max (maximum specific growth rate), lag time, and maximum population. To characterize the influence of temperature on the maximum specific growth rate and lag phase length, the square root model (secondary model) may be used.

The models are fitted to *Pseudomonas* spp. growth data, and all parameters may be determined using an appropriate fitting tool, such as the Matlab 8.3.0.532 (R2014a) software's NonLinearModel command, which employs the Levenberg Marquardt technique (MathWorks Inc., Natick, MA, USA) (*see* **Note 1**). The basic entries in Matlab are shown in Fig. 2. The basic entries are time, microbial concentration, and temperature.

1. For predicting growth parameters, the Baranyi and square root models follow the script in Matlab (*see* **Note 2**).

```
%This Code was written for Baranyi model.
%nmax was calculated basing on 1/h.
%microbial load was fitted as log10 CFU/g.
%%
clc;
clear all;
%% %%%%%%%%%%%%%%% EXPERIMENTAL DATA %%%%%%%%%%%%%%%%%%%%
for i=1:4;
```

```
if i== 1;exp_data=importdata('4_C.txt');else; end;
if i== 2;exp_data=importdata('8_C.txt');else; end;
if i== 3;exp_data=importdata('12_C.txt');else; end;
if i== 4;exp_data=importdata('16_C.txt');else; end;

% %
.........................................................-
..
t=(exp_data.data(:,1)); % Time data
data_log=(exp_data.data(:,2)); % Concentration log
T=(exp_data.data(:,3)); % Temperature
%% %%%%%%%%% GUESS of PARAMETERS for ONE STEP METHOD %%%%%%%%%%
%%%%%
%%
ymin0=(min(data_log)); %Parameter-1
ymax0=(max(data_log)); %Parameter-2
for k=2:length(data_log);
 a(k-1)=((data_log(k)- data_log(k-1))/(t(k)-t(k-1)));
end
nmax=max(a)*log(10); %Parameter-3
lamda=t(end)/4; %Parameter-4
guess=[ymin0 ymax0 nmax lamda];
myFit_wl = NonLinearModel.fit(t,data_log,@f_Baranyi_two_step,
guess);
%% CONTROL of if There is Lag Phase
if myFit_wl.Coefficients.Estimate(4)>0;
 parameters(:,i)=myFit_wl.Coefficients.Estimate(1:2);
 SE_parameters(:,i)=myFit_wl.Coefficients.SE(1:2);
msgr(:,i)=myFit_wl.Coefficients.Estimate(3);
SE_msgr(:,i)=myFit_wl.Coefficients.SE(3);
lag(:,i)=myFit_wl.Coefficients.Estimate(4);
SE_lag(:,i)=myFit_wl.Coefficients.SE(4);end;
if myFit_wl.Coefficients.Estimate(4)<0;
display('There was no lag phase observed')
no_lag=myFit_wl.Coefficients.Estimate(1:3)';
myFit_nl = NonLinearModel.fit(t,data_log,@f_Baranyi_two_s-
tep_no_lag, no_lag);
msgr(:,i)=myFit_nl.Coefficients.Estimate(3);
SE_msgr(:,i)=myFit_nl.Coefficients.SE(3);
lag(:,i)=0;SE_lag(:,i)=0;
 parameters(:,i)=myFit_nl.Coefficients.Estimate(1:2);
 SE_parameters(:,i)=myFit_nl.Coefficients.SE(1:2);
% myFit = NonLinearModel.fit(t,data_log,@f_Baranyi_two_step,
guess);
% %% CONTROL of if There is Lag Phase
% if myFit.Coefficients.Estimate(4)<25;
% display('There was no lag phase observed')
% no_lag=myFit.Coefficients.Estimate(1:3)';
```

```
% myFit = NonLinearModel.fit(t,data_log,@f_Baranyi_two_step_-
no_lag, no_lag)
% parameters(:,i)=myFit.Coefficients.Estimate;
% SE_parameters(:,i)=myFit.Coefficients.SE;
% end
end
Temperature(i)=unique(T);

RESULTS(:,i)=[myFit_wl.Coefficients.Estimate(1:4);myFit_wl.
Coefficients.SE(1:4);myFit_wl.MSE;myFit_wl.Rsquared.Ordinary]

end

RESULTS=RESULTS'
msgr

%% RATKOWSYKY-MAXIMUM-GROWTH-RATE
n_max=msgr;
growthsqr=sqrt(n_max);
[a,b]=polyfit(Temperature,growthsqr,1);
guessRn=[a(1) 0];
myFit = NonLinearModel.fit(Temperature,n_max,@f_Ratkowsky_n-
max,guessRn);
parameters_R_nmax=myFit.Coefficients.Estimate;
SE_parameters_R_nmax=myFit.Coefficients.SE;

%% RATKOWSYKY-LAMDA
lamda=lag(2:4);
guessRl=[0];
myFit = NonLinearModel.fit(n_max(2:4),lamda,@f_Ratkowsky_lam-
da,guessRl);
parameters_R_lamda=myFit.Coefficients.Estimate;
SE_parameters_R_lamda=myFit.Coefficients.SE;

PARAMETERS=[mean(parameters(1,:));mean(parameters(2,:));para-
meters_R_nmax;parameters_R_lamda]';
SE_PARAMETERS=[mean(SE_parameters(1,:));mean(SE_parameters
(2,:));SE_parameters_R_nmax;SE_parameters_R_lamda]';

%% prediction of microbial population
exp_data=importdata('Milk.txt');
OBSERVED=(exp_data.data(:,2));
t=(exp_data.data(:,1)); % Time
T=(exp_data.data(:,3)); % Temperature
tT=[t,T];
PREDICTION=f_Baranyi_two_step_prediction([PARAMETERS],tT);
```

```
t1_r=linspace(0,1370,100);
t2_r=linspace(0,790,100);
t3_r=linspace(0,259,100);
t4_r=linspace(0,187,100);
t5_r=linspace(0,141,100);
treshape=[t1_r t2_r t3_r t4_r t5_r]';
Treshape=[ones(100,1)*1.5; ones(100,1)*4; ones(100,1)*8; ones
(100,1)*12; ones(100,1)*16];
tTreshape=[treshape,Treshape];
PREDICTION_re=f_Baranyi_two_step_prediction([PARAMETERS],
tTreshape);
%%
Tmin=min(T);Tmax=max(T);l=100;
temps=linspace(Tmin,Tmax,l);
Tmin_for_nmax=parameters_R_nmax(1);
b1_for_nmax=parameters_R_nmax(2);
adjustment=parameters_R_lamda;
for m=1:length(temps)
Ti=temps(m);
nmax(m)=((b1_for_nmax*(Ti-Tmin_for_nmax)).^2);
lamda(m)=adjustment/(nmax(m));
end

SECONDARY_MODEL_RESULTS=[temps' nmax' lamda'];

%% Evaluation Goodness of Fitting
n=length(OBSERVED);s=5;ssR=sum((PREDICTION-OBSERVED).^2);
mSE=ssR/n; ssTOT=sum((OBSERVED-mean(OBSERVED)).^2);
R2_ADJUSTED=1-(((n-1)./(n-s)).*(ssR/ssTOT));
RMSE=sqrt(ssR/(n-s));
BIC=n*log(ssR/n)+s*log(n)

AICc=n*log(ssR/n)+ (2*(s+1)) + ((2*(s+1)*(s+2))/(n-s-2))

RESULTS=[mean(parameters(1,:));mean(parameters(1,:))-mean
(SE_parameters(1,:));mean(parameters(1,:))+mean(SE_parameters
(1,:)); mean(parameters(2,:)); (mean(parameters(2,:))-mean
(SE_parameters(2,:)));(mean(parameters(2,:))+mean(SE_para-
meters(2,:))); parameters_R_nmax(1); (parameters_R_nmax(1)-
SE_parameters_R_nmax(1));(parameters_R_nmax(1)+SE_parame-
ters_R_nmax(1));parameters_R_nmax(2); (parameters_R_nmax
(2)-SE_parameters_R_nmax(2));(parameters_R_nmax(2)+SE_parame-
ters_R_nmax(2)); parameters_R_lamda; (parameters_R_lamda-SE_-
parameters_R_lamda);(parameters_R_lamda
+SE_parameters_R_lamda); RMSE;R2_ADJUSTED; AICc]'

[RMSE;R2_ADJUSTED; AICc; BIC]'
```

5 Comparison of the Goodness of Fit of the Models

The root mean square error (RMSE) and adjusted coefficient of determination ($R^2{}_{\text{adj}}$), Akaike information criteria (AIC), and Bayesian information criterion (BIC) may be used to compare the models' estimate skills using Eqs. 7, 8, 9, and 10 correspondingly [27]:

$$\text{RMSE} = \sqrt{\sum_{i=1}^{n} \frac{\left(x_{\text{obs}} - x_{fit}\right)^2}{\text{n} - \text{s}}} \tag{7}$$

$$R^2_{\text{adj}} = 1 - \left(\frac{n-1}{n-s}\right)\left(\frac{\text{SSE}}{\text{SST}}\right) \tag{8}$$

$$\text{AIC} = n \ln\left(\frac{\text{SSE}}{n}\right) + 2s \tag{9}$$

$$\text{BIC} = n \ln\left(\frac{\text{SSE}}{n}\right) + s \ln(n) \tag{10}$$

where x_{obs} represents the experimental bacterial growth concentration, x_{fit} represents the fitted value, n represents the number of experiments, s represents the number of model parameters, SSE represents the sum of squares of errors, and SST represents the total sum of squares.

Because the major models employ different scales for counting microbe populations, the modified Gompertz and logistic models' RMSE, AIC, and BIC values cannot be linked directly to the Baranyi and Huang models' RMSE, AIC, and BIC values. As a result, the required conversion from the ln scale to the log scale or from the log scale to the ln scale should be carried out.

6 The Models' Validation

The validation of models is the process through which the predictive power of constructed models is validated using previously published or newly generated data. The ability of the models to forecast may be assessed using the microbes' growth kinetics. Each of the global models' related bias (B_f) and accuracy (A_f) factors [28] are presented in Eqs. 11 and 12, respectively, for comparison:

$$B_{\text{f}} = 10^{\frac{\sum_{i=1}^{n} \log\left(\frac{x_{\text{pred}}}{x_{\text{obs}}}\right)}{\text{n}}} \tag{11}$$

$$A_f = 10^{\frac{\sum_{i=1}^{n} \left|\log\left(x_{\text{pred}}/x_{\text{obs}}\right)\right|}{n}} \tag{12}$$

where x_{pred} denotes projected maximum values (1/h) and (h), x_{obs} denotes experimental μ_{max} (1/h) and λ (h), and n is the number of experimental growth data.

7 Notes

1. In a nonlinear regression technique, determining initial values is a critical step in estimating appropriate parameters. The smallest and maximum concentrations of microbe populations for all temperature ranges, respectively, were chosen as starting values for the parameters x_0 and x_{max}. Because choosing any starting points for these empiric parameters could lead the estimated parameters to possible local optimal points around the global one, the starting values of b_1 and T_0 for the maximum growth rate and b_2 for the lag phase duration can be estimated using the ga command in the Global Optimization Toolbox of Matlab software.
2. To run the code for primary and secondary models in Matlab, all files called in the script need to be saved in the .txt extension. Also, the file names should not contain any special characters such as!#@$%^&*() − +°.

References

1. Hammond ST, Brown JH, Burger JR et al (2015) Food spoilage, storage, and transport: implications for a sustainable future. Bioscience 65(8):758–768
2. Burkepile DE, Parker JD, Woodson CB et al (2006) Chemically-mediated competition between microbes and animals: microbes as consumers in food webs. Ecology 87:2821–2831
3. Braun PG (2006) In: Blackburn CW (ed) Modelling microbial food spoilage. Woodhead Publishing, Cambridge
4. Bovill RA, Bew J, Baranyi J (2001) Measurements and predictions of growth for *Listeria monocytogenes* and *Salmonella* during fluctuating temperature. II. Rapidly changing temperatures. Int J Food Microbiol 67(1–2):131–137. https://doi.org/10.1016/s0168-1605(01)00446-9
5. Rawat S (2015) Food spoilage: microorganisms and their prevention. Asian J Plant Sci 5(4):47–56
6. Adams MR, Moss MO (2000) Food microbiology. Royal Society of Chemistry, Cambridge
7. Bezirtzoglou E, Maipa V, Voidarou C et al (2009) Food-borne intestinal bacterial pathogens. Anaerobe 12(2):96–104. https://doi.org/10.3402/mehd.v12i2.8073
8. Stavropoulou E, Bezirtzoglou E (2019) Predictive modeling of microbial behavior in food. Foods 8(12):654. https://doi.org/10.3390/foods8120654
9. Kantor LS, Upton K, Manchester A et al (1997) Estimating and addressing America's food losses. Food Rev 20:2–12
10. WWF-UK (2021) https://wwfeu.awsassets.panda.org/downloads/driven_to_waste___the_global_impact_of_food_loss_and_waste_on_farms.pdf. Accessed 15 Aug 2021
11. Mekonnen MM, Gerbens-Leenes W (2020) The water footprint of global food production. Water 12(10):2696. https://doi.org/10.3390/w12102696
12. Per'ez-Rodrı'guez F, Valero A (2013) Predictive microbiology in foods. Springer, New York
13. Fujikawa H (2016) Prediction of competitive microbial growth. Biocontrol Sci 21(4):215–223. https://doi.org/10.4265/bio.21.215
14. Raposo A, Pérez E, Tinoco de Faria C, Carrascosa et al (2016) Food spoilage by *pseudomonas* spp.—an overview. In: Singh OV, John Wiley & Sons (eds) Foodborne pathogens and antibiotic resistance. John Wiley & Sons Inc, Hoboken
15. Koutsoumanis K (2001) Predictive modeling of the shelf life of fish under nonisothermal conditions. Appl Environ Microbiol 67(4):1821–1829
16. Gospavic R, Kreyenschmidt J, Bruckner S et al (2008) Mathematical modelling for predicting the growth of *Pseudomonas* spp. in poultry under variable temperature conditions. Int J Food Microbiol 127(3):290–297

17. Zhang L, Hinz AJ, Nadeau JP et al (2011) *Pseudomonas aeruginosa* tssC1 links type VI secretion and biofilm-specific antibiotic resistance. J Bacteriol 193(19):5510–5513
18. Bruckner S, Albrecht A, Petersen B et al (2013) A predictive shelf life model as a tool for the improvement of quality management in pork and poultry chains. Food Control 29(2): 451–460
19. Dabadé DS, den Besten HMW, Azokpota P et al (2015) Spoilage evaluation, shelf-life prediction, and potential spoilage organisms of tropical brackish water shrimp (*Penaeus notialis*) at different storage temperatures. Food Microbiol 48:8–16
20. Lytou A, Panagou EZ, Nychas GJE (2016) Development of a predictive model for the growth kinetics of aerobic microbial population on pomegranate marinat ed chicken breast fillets under isothermal and dynamic temperature conditions. Food Microbiol 55:25–31
21. Simon A, Gonzalez-Fandos E, Tobar V (2005) The sensory and microbiological quality of fresh sliced mushroom (*Agaricus bisporus L.*) packaged in modified atmospheres. Int J Food Sci 40(9):943–952. https://doi.org/10.1111/j.1365-2621.2005.01028.x
22. Heir E, Moen B, Åsli AW et al (2021) Antibiotic resistance and phylogeny of *pseudomonas* spp. isolated over three decades from chicken meat in the Norwegian food chain. Microorganisms 9(2):207
23. Tarlak F, Ozdemir M, Melikoglu M (2020) Predictive modelling for the growth kinetics of *Pseudomonas* spp. on button mushroom (*Agaricus bisporus*) under isothermal and non-isothermal conditions. Food Res Int 130: 108912
24. Whiting RC (1995) Microbial modeling in foods. Crit Rev Food Sci Nutr 35(6):467–494
25. Ratkowsky DA, Olley J, McMeekin TA et al (1982) Relationship between temperature and growth rate of bacterial cultures. J Bacteriol 149(1):1–5. https://doi.org/10.1128/jb.149.1.1-5.1982
26. Wang J, Rahman SME, Zhao XH et al (2013) Predictive models for the growth kinetics of *Listeria monocytogenes* on white cabbage. J Food Saf 33(1):50–58
27. Taoukis PS, Koutsoumanis K, Nychas GJE (1999) Use of time–temperature integrators and predictive modelling for shelf life control of chilled fish under dynamic storage conditions. Int J Food Microbiol 53:21–31
28. Juneja VK, Melendres MV, Huang L et al (2007) Modeling the effect of temperature on growth of *salmonella* in chicken. Food Microbiol 24:328–335

Chapter 12

Quantitative PCR Technique for Predictive Models

Wiaslan Figueiredo Martins and Mirian Cristina Feiten

Abstract

This chapter reports the use of real-time quantitative polymerase chain reaction (qPCR) to quantify *Weissella viridescens*, a spoilage bacteria of meat products that cause sour odors, off-flavors, vacuum loss, white exudate production, and slime production on vacuum-packed cooked sausage. This molecular approach represents a reliable and sensitive tool to detect and quantify spoilage bacteria in DNA extracted from foods, and its use can also be recommended to study bacterial behavior in meat products by quantifying bacterial growth in the latent phase, when signs of deterioration are not present yet, up to the growth stationary phase. Also, molecular predictive models from the qPCR data can be established to predict the meat products' shelf life.

Key words Bacteria quantification, qPCR, Meat products' shelf life, Spoilage, Predictive modeling

1 Introduction

The microbiological quality should be effectively focused on traceability, with emphasis on the "farm-to-fork" approach, reaching the entire food chain. For that, rapid diagnostic methods are highly recommended so that control strategies can be applied, aiming at ensuring the consumer's health and reducing food waste, therefore, preventing significant financial losses to the industry [1].

Meat products stored under refrigeration temperatures can be spoiled mainly by psychrotrophic lactic acid bacteria (LAB), along with other microbial groups and species. They are known as mixed communities of species that dominate the general bacterial population after a few days or weeks [2]. LAB may be the cause and are considered agents of meat spoilage, like the genera *Weissella* and *Leuconostoc*, associated with a wide range of meat products, including vacuum-packed fresh meat and processed meat [3, 4].

Rapid diagnostic methods are essential for the assessment of food safety objectives (minimum levels of hazards at the moment of consumption) in food safety management, which require results in a shorter time than those obtained by culture cultivation [5]. In this

Verônica Ortiz Alvarenga (ed.), *Basic Protocols in Predictive Food Microbiology*, Methods and Protocols in Food Science,
https://doi.org/10.1007/978-1-0716-3413-4_12,

sense, the application of quantitative polymerase chain reaction (qPCR) has become the molecular method of choice for the absolute quantification of spoilage microorganisms in recent years [6–9]. The qPCR has been used to measure the growth kinetics of foodborne pathogens, such as *Salmonella enteritidis* in chicken juice [10], *Listeria monocytogenes* in vacuum-packed pork meat [11], *Staphylococcus aureus* in pork meat [12], *Salmonella* spp. in pasteurized and non-pasteurized milk [13], and *Carnobacterium maltaromaticum*, *Brochothrix thermosphacta*, and *Serratia liquefaciens* in mixed culture [14].

2 Materials

2.1 Inoculum Preparation

1. *Weissella viridescens* strain from reference culture collection or food isolates.
2. De Man, Rugosa, and Sharpe (MRS) broth.
3. Glycerol.
4. Ethanol (70%).
5. Sterile microcentrifuge tubes (2 mL) RNase- and DNase-free.
6. Disposable gloves.
7. Sterile borosilicate glass culture tube (6 × 150 mm).
8. Pipettes and sterile disposable pipette tips.
9. Vortex mixer.
10. Spectrophotometer (e.g., BEL Engineering).
11. Spectrophotometer cuvettes.
12. Bacteriological culture greenhouse.

2.2 Meat Product Sample Preparation

1. Meat product (e.g., blood sausages).
2. Inoculum *W. viridescens* strain (*see* Subheading 3.1).
3. Plastic bags.
4. Pipettes and sterile pipette tips.
5. Scissors.
6. Thermostatic stirring bath.
7. Ethanol (70%).
8. Sterile distilled water.
9. Disposable gloves.

2.3 Quantification by the Standard Plate Count Method

1. Sterile plastic bags with a filter.
2. Sterile Petri dishes (plastic or glass).
3. Pipettes and sterile disposable pipette tips.
4. Ringer's solution.

5. Stomacher homogenizer.
6. MRS agar.
7. Bacteriological culture greenhouse.

2.4 Genomic DNA Extraction from Blood Sausage

1. Bacterial genomic DNA kit (e.g., GenElute™ Kit, Sigma-Aldrich, USA) (*see* **Note 1**).
2. Sterile centrifuge tubes (50 mL).
3. Centrifuge (e.g., Centrifuge 5810R, Eppendorf®).
4. Sterile microcentrifuge tubes (1.5 mL) RNase- and DNase-free.
5. Micropipettes and sterile disposable pipette tips.
6. Spectrophotometer (e.g., NanoDrop).
7. Microcentrifuge.
8. Vortex mixer.
9. Disposable gloves.

2.5 qPCR Assay

1. qPCR instrument (e.g., StepOne™ Real-Time PCR System, Applied Biosystems).
2. Optical eight-tube strip (e.g., MicroAMP®, Applied Biosystems).
3. Primers and TaqMan probe (*see* Subheading 3.5.2, **step 1**).
4. Universal PCR master mix (e.g., FastStart™ Universal Probe Master (ROX), Roche Molecular Systems).
5. Nuclease-free water.
6. Pipettes and sterile disposable pipette tips.
7. Disposable gloves.

3 Methods

3.1 Inoculum Preparation

1. For inoculum preparation, 100 μL of *W. viridescens* strain suspension is used in MRS broth and supplemented with 30% glycerol.
2. The strain suspension is transferred to a borosilicate glass tube containing 10 mL of MRS broth. It is then incubated at 30 °C for 24 h and subcultured on the following day. The subculture is grown for 18 h until a stationary growth phase ensues (10^9 CFU/mL) (*see* **Note 2**).

3.2 Meat Product Sample Preparation

1. Sliced blood sausage samples are vacuum-packed and pasteurized in a thermostatic bath at 75 °C for 10 min according to Diez et al. [15] (*see* **Note 3**).

2. After the pasteurization treatment, each sample is opened and artificially inoculated with 200 μL of inoculum containing approximately 10^5 CFU/mL (*see* **Note 4**) of *W. viridescens* and vacuum-packed again.
3. Nuclease-free water (200 μL) is added to the non-inoculated control samples (negative control) (*see* **Note 5**).
4. Samples are incubated at the temperatures at which the shelf life of the meat is to be studied (e.g., product temperature, temperatures that simulate an interruption of the cold chain, or optimal bacterial growth). At each sample of the growth curve, 2 mL of the supernatant is taken in duplicate for qPCR analysis (*see* **Note 6**).

3.3 Growth Curve Construction

Artificially inoculated meat product samples (e.g., blood sausage) are isothermally incubated at different temperatures (e.g., temperatures to reflect the different storage conditions of refrigerated foods). Microbial growth is measured in duplicate with both the qPCR and the plate count method until the stationary growth phase.

3.4 Quantification of W. viridescens by Plate Count Method

Decimal dilutions of Ringer's solution are prepared for microbiological analysis, and 100 μL aliquots of the appropriate dilutions are plated on MRS agar in duplicate at 30 °C for 48 h, using the method specified by the International Organization for Standardization – ISO 15214:1998 [16]. Non-inoculated control samples are analyzed throughout the experiments. A mean CFU/g (and deviation) of two independent samples of each growth curve point is calculated and used to determine the growth kinetics. The initial count (day 0) of strains in the meat product is determined by plate count on MRS after inoculation.

3.5 qPCR Assay

There are different types of DNA extraction protocols available in the literature, and the choice of the protocol should be based on the best DNA yield for the type of material to be processed (e.g., culture medium or food). For short-term storage of the DNA, 2–8 °C is recommended. For longer-term storage, −20 °C is recommended. Avoid freezing and thawing, which causes breaks in the DNA strand. In this protocol, the GenElute™ Bacterial Genomic DNA kit protocol is used, following the manufacturer's recommendations for Gram-positive bacteria.

3.5.1 Genomic DNA Extraction from Blood Sausage

1. The first step is the preparation of a 2.115×10^6 unit/mL stock solution of lysozyme, as described by the manufacturer. For each DNA preparation to be performed, 200 μL of lysozyme solution is required (*see* **Note 7**).
2. Then, extract the DNA from blood sausage inoculated with *W. viridescens* by using a cell suspension of 40 mL of homogenate, which is then centrifuged at 300 × *g* for 10 min (4 °C).

3. A volume of 1.5 mL of the supernatant is subsequently collected by centrifugation at 5000 × *g* for 15 min (4 °C), and their genomic DNA is isolated using the GenElute™ kit protocol for Gram-positive bacteria (*see* **Note 8**).
4. Next, pellet 1.5 mL of an overnight bacterial broth culture by centrifuging for 2 min at 12,000 × *g*. Remove the culture medium completely and discard it.
5. Resuspend the pellet thoroughly in 200 μL of lysozyme solution (*see* Subheading 3.5.1, **step 1**), and incubate for 30 min at 37 °C.
6. Add 20 μL RNase A solution and incubate for 2 min at room temperature.
7. Add 20 μL of the proteinase K solution to the sample, followed by 200 μL of lysis solution C.
8. Vortex thoroughly (about 15 s) and incubate at 55 °C for 10 min (*see* **Note 9**).
9. Add 500 μL of Column Preparation Solution to the binding column previously placed into a 2 mL collection tube, without wetting the circlet. Close the cap and centrifuge at 12,000 × *g* for 1 min (*see* **Note 10**). Discard the flow-through liquid but retain the collection tube.
10. Add 200 μL of ethanol (95–100%) to the lysate from step 6, and mix thoroughly by vortexing for 10 s. A homogeneous mixture is essential.
11. Transfer the entire contents of the tube into the treated binding column from step 9. Use a wide-bore pipette tip to reduce the DNA shearing when transferring contents into the binding column. Centrifuge at minimum 6500 × *g* for 1 min. Replace the column with a new 2 mL collection tube, and discard the previous tube containing the flow-through liquid.
12. Add 500 μL of Column Wash Solution 1 to the column, and centrifuge for 1 min at 6500 × *g*. Discard the flow-through liquid but retain the collection tube.
13. Add 500 μL of Wash Solution Concentrate, previously diluted with ethanol, to the column, and centrifuge for 3 min at maximum speed (16,000 × *g*) to dry the column (*see* **Note 11**).
14. Discard the collection tube containing the flow-through liquid, and place the binding column in a new 2 mL collection tube. Carefully add 200 μL of the Elution Solution directly into the center of the column, and incubate for 5 min at room temperature.
15. Centrifuge for 1 min at 6500 × *g* to elute the DNA (*see* **Note 12**).

3.5.2 Quantification of W. viridescens by qPCR

1. Absolute quantification of *W. viridescens* DNA is carried out by amplifying the target region of small subunits of *recN* gene by using the following primers and TaqMan probe:

   Forward primer: 5'–CGCAAACACAACAAGCCTAT–3'

   Reverse primer: 5'–TGTTGAGCAAGTTCCAAAGC–3'

   TaqMan probe: 5'–FAM- CCGTGCCCTGACGTTAGCCA - BHQ1–3'
2. *W. viridescens* DNA is obtained from the blood sausage inoculated (*see* Subheading 3.2), and the concentrations are measured at 260 nm using a NanoDrop spectrophotometer.
3. The standard curve is generated by using an eightfold serial dilution (ranging from 10^1 to 10^8 CFU/g).
4. For each run, include the suitable control samples and NTC (No Template Control).
5. Prepare the qPCR reaction in a final volume of 25 μL by using the following:
   (a) 300 nM forward primer
   (b) 300 nM reverse primer
   (c) 175 nM TaqMan probe
   (d) 12.5 μL FastStart Universal Probe Master (ROX)
6. Add 2 μL of template DNA.
7. Run each sample in triplicate, according to the following qPCR protocol: 95 °C (5 min); 45 cycles of 95 °C (1 min); 64 °C (0.5 min); 72 °C (0.5 min); and a final extension for 10 min at 72 °C.
8. The unknown sample concentrations will be calculated by interpolation of the data on the standard curve (Fig. 1).
9. For each sample on the growth curve, the amount of *W. viridescens* DNA inoculated is expressed as $\log_{10}$ CFU/g (Fig. 2).
10. Different models can be fitted to the growth data obtained by qPCR and the plate count method to measure lag phase duration, growth rate, and maximum population density (*see* **Note 13**).

Figure 1 shows the standard curve obtained for a cocktail of *Weissella viridescens* (WV) in vacuum-packed *Morcilla de Burgos* (VPM), while Fig. 2 depicts growth curves of *Weissella viridescens* (WV) at different storage temperatures.

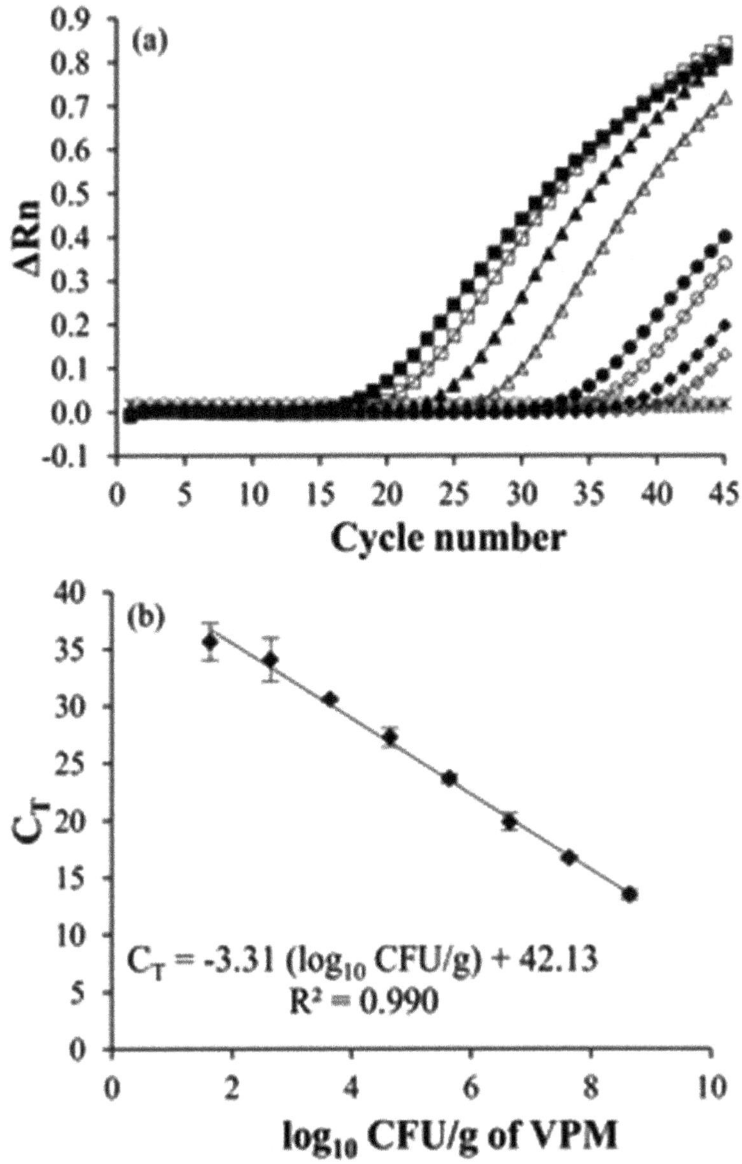

Fig. 1 Standard curve obtained for a cocktail of *Weissella viridescens* (WV) in vacuum-packed *Morcilla de Burgos* (VPM). Representative amplification plots. (**a**) Increase in fluorescence (ΔRn) with respect to the number of qPCR cycles for serial dilutions of a cocktail of WV corresponding to 4.3×10^8 (-■-), 4.3×10^7 (-□-), 4.3×10^6 (-▲-), 4.3×10^5 (-Δ-), 4.3×10^4 (-●-), 4.3×10^3 (-○-), 4.3×10^2 (--), 4.3×10^1 (-◇-) CFU/g, non-inoculated control sample (–), and threshold (-x-). (**b**) Standard curve constructed by plotting the mean C_T values for six qPCR runs for two independent amplifications against the logarithm of the cell concentration in CFU (determined by plate counting) of WV. The error bars show the standard deviation. (Reproduced from Ref. [17] with permission from Elsevier)

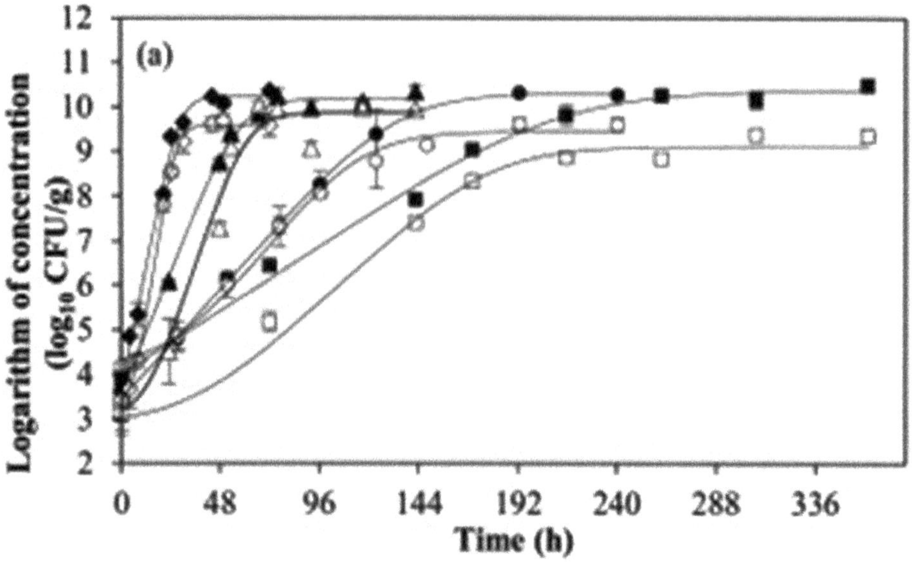

Fig. 2 Growth curves of *Weissella viridescens* (WV) at 5 °C (square symbols), 8 °C (circle symbols), 13 °C (triangle symbols), and 18 °C (lozenge symbols), quantified by plate count (PC, filled symbols) and qPCR (unfilled symbols) methods, and the fitting of Baranyi and Roberts model (BAR, continuous lines) to the experimental data obtained in vacuum-packed *Morcilla de Burgos* (VPM). The error bars show the standard deviation. (Reproduced from Ref. [17] with permission from Elsevier)

4 Notes

1. DNA kits for food can also be used.
2. The subculture is grown for 18 h until a stationary growth phase ensues. At that time, the cell concentration is in the stationary phase, confirmed by the optical density. In this case, the value of 0.9 measured at 600 nm, in a 1:10 dilution, indicates a concentration of 10^9 CFU/mL.
3. The weight of the slices was standardized (e.g., 20 g) and packed in the shake bag for storage. Thus, the microbiological analysis was facilitated since the same amount of water was added to each sample. In addition, the slices were pasteurized in plastic vacuumed bags to eliminate the natural microbiota.
4. The initial concentration of 10^3 CFU/g simulates the initial microbiota of meat products, making the results more realistic.
5. As 200 μL of bacteria in broth are added, the non-inoculated control sample is added to 200 μL of nuclease-free water so that the amount of liquids is standardized in the inoculated and control samples.
6. Consider collecting data for qPCR analysis until the beginning of the stationary phase, avoiding dead cells.

7. Prepare an extra solution in case of pipetting errors.
8. For foods such as blood sausage, rice, etc., a filter is important to avoid pieces of food that hinder DNA quality.
9. To assure an efficient lysis process, it is fundamental that the sample is narrowly mixed for 15 s by pulse vortexing.
10. After centrifugation, if the solution has not entirely passed through the membrane, repeat the centrifugation step at a higher speed until all the solution has passed through.
11. The column must be free of ethanol before eluting the DNA. Centrifuge the column for an additional 1 min at maximum speed if residual ethanol is seen.
12. Repeat this step twice.
13. Different models described by sigmoid functions can be fitted to the growth data to obtain the kinetic parameters: modified Gompertz and logistic (Baranyi and Roberts), Zwietering et al. and Huang [18–20]. To fit the Baranyi and Roberts model, DMFit v3.5 (ComBase; http://www.combase.cc/tools/) software can be used, free of charge.

References

1. Lopes ATS, Maciel MB (2020) Real-time quantitative PCR as a tool for monitoring microbiological quality of food. In: Nagpal ML, Boldura O, Baltă C, Enany S (eds) Synthetic biology – new interdisciplinary science. IntechOpen, London. https://doi.org/10.5772/intechopen.84532
2. Lucquin I, Zagorec M, Champomier-Vergès M, Chaillou S (2012) Fingerprint of lactic acid bacteria population in beef carpaccio is influenced by storage process and seasonal changes. Food Microbiol 29:187–196. https://doi.org/10.1016/j.fm.2011.08.001
3. Han Y, Xu X, Jiang Y, Zhou G, Sun X, Xu B (2010) Inactivation of food spoilage bacteria by high pressure processing: evaluation with conventional media and PCR–DGGE analysis. Food Res Int 43:1719–1724. https://doi.org/10.1016/j.foodres.2010.05.012
4. Diez AM, Santos EM, Jaime I, Rovira J (2009a) Effectiveness of combined preservation methods to extend the shelf life of *Morcilla de Burgos*. Meat Sci 81:171–177. https://doi.org/10.1016/j.meatsci.2008.07.015
5. Santana AS, Franco BDG (2009) Microbial quantitative risk assessment of foods: concepts, systematics and applications. Braz J Food Technol 12:266. https://doi.org/10.4260/BJFT2009800900021
6. Gómez-Rojo EM, Romero-Santacreu L, Jaime I, Rovira J (2015) A novel real-time PCR assay for the specific identification and quantification of *Weissella viridescens* in blood sausages. Int J Food Microbiol 215:16. https://doi.org/10.1016/j.ijfoodmicro.2015.08.002
7. Martínez N, Martín MC, Herrero A et al (2011) qPCR as a powerful tool for microbial food spoilage quantification: significance for food quality. Trends Food Sci Technol 22:367. https://doi.org/10.1016/j.tifs.2011.04.004
8. Elizaquível P, Chenoll E, Aznar A (2008) A TaqMan-based real-time PCR assay for the specific detection and quantification of Leuconostoc mesenteroides in meat products. FEMS Microbiol Lett 278:62. https://doi.org/10.1111/j.1574-6968.2007.00974.x
9. Martín B, Jofré A, Garriga M et al (2006) Rapid quantitative detection of *lactobacillus sakei* in meat and fermented sausages by real-time PCR. Appl Environ Microbiol 72:6040. https://doi.org/10.1128/AEM.02852-05
10. Noviyanti F, Hosotani Y, Koseki S, Inatsu Y, Kawasaki S (2018) Predictive modeling for the growth of *salmonella* Enteritidis in chicken juice by real-time polymerase chain reaction. Foodborne Pathog Dis 15:406–412

11. Ye K, Wang H, Zhang X, Jiang Y, Xu X, Zhou G (2013) Development and validation of a molecular predictive model to describe the growth of *listeria monocytogenes* in vacuum-packaged chilled pork. Food Control 32:246–254
12. Guan Z, Jiang Y, Gao F, Zhang L, Zhou G, Guan Z (2017) Analysis of a molecular predictive mode for the growth of *Staphylococcus aureus* in pork. Int J Food Prop 20:68–82
13. Kawasaki S, Shimizu S, Koseki S, Inatsu Y (2014) Characterization and modeling of *salmonella* growth in pasteurized and nonpasteurized milk using real-time PCR. Jap J Food Microbiol 31:28–35
14. Mohsina K, Kaur M, Bowman JP, Powell S, Tamplin ML (2020) qPCR quantification of *Carnobacterium maltaromaticum*, *Brochothrix thermosphacta*, and *Serratia liquefaciens* growth kinetics in mixed culture. J Microbiol Methods 175:105961. https://doi.org/10.1016/j.mimet.2020.105961
15. Diez AM, Jaime I, Rovira J (2009b) The influence of different preservation methods on spoilage bacteria populations inoculated in morcilla de Burgos during anaerobic cold storage. Int J Food Microbiol 132:91–99
16. Anonymous (1998) Microbiology of food and animal feeding stuffs – horizontal method for the enumeration of mesophilic lactic acid bacteria – colony count technique at 30 °C. BS ISO 14214. BSI, London, UK
17. Martins WF, Longhi DA, Aragão GMF et al (2020) A mathematical modeling approach to the quantification of lactic acid bacteria in vacuum-packaged samples of cooked meat: combining the TaqMan-based quantitative PCR method with the plate-count method. Int J Food Microbiol 318:108466. https://doi.org/10.1016/j.ijfoodmicro.2019.108466
18. Baranyi J, Roberts T (1994) A dynamic approach to predicting bacterial growth in food. Int J Food Microbiol 23:277–294
19. Huang LH (2004) Numerical analysis of the growth of *Clostridium perfringens* in cooked beef under isothermal and dynamic conditions. J Food Saf 24:53–70
20. Zwietering MH, Jongenburger I, Rombouts FM, Van't RK (1990) Modeling of the bacterial growth curve. Appl Environ Microbiol 56:1875–1881

Index

Verônica Ortiz Alvarenga (ed.), *Basic Protocols in Predictive Food Microbiology*, Methods and Protocols in Food Science, https://doi.org/10.1007/978-1-0716-3413-4,

Printed in the USA
CPSIA information can be obtained
at www.ICGtesting.com
LVHW012041130923
758136LV00007B/68

9 781071 634127